KNOWLEDGE ENCYCLOPEDIA
HUMAN BODY
HEART & CIRCULATORY SYSTEM

© Wonder House Books 2024

All rights reserved. No part of this book may be reproduced or transmitted in any form by any means, electronic or mechanical, including photocopying and recording, or by any information storage and retrieval system except as may be expressly permitted in writing by the publisher.

(An imprint of Prakash Books)

contact@wonderhousebooks.com

Disclaimer: The information contained in this encyclopedia has been collated with inputs from subject experts. All information contained herein is true to the best of the Publisher's knowledge.

ISBN : 9789389931204

Table of Contents

The Amazing Human Body	3
At the Heart of Everything	4–5
Inside Your Heart	6–7
Arteries & Veins	8–9
Feeding the Organs: Capillaries	10
Keeping the Beat	11
Heart Troubles	12–13
Transplants: Gifting A New Life	14
Know Your Blood	15
Blood Cells: Little Transporters & Defenders	16–17
Platelets & Plasma	18
Donate Blood, Save a Life	19
Know Your Blood Type	20–21
Know Your Blood Pressure	22
Lymph: The Second Circulatory System	23
Lymphatic System: Your Body's Janitor	24–25
Blood Disorders	26–27
Life Cycle of Blood Cells	28
Watch Out	29
A Healthy Heart Makes a Healthy Body	30-31
Word Check	32

THE AMAZING HUMAN BODY

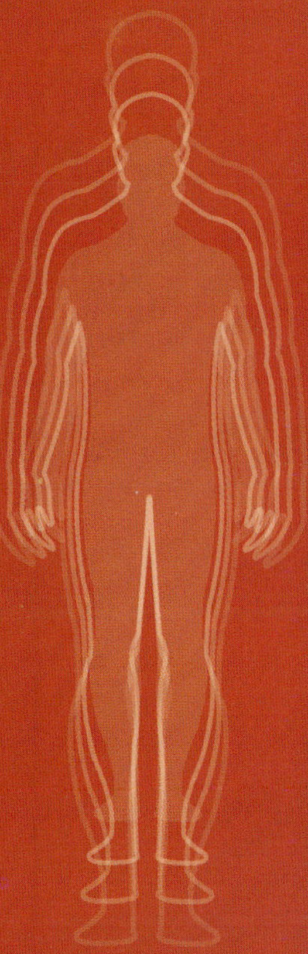

You might not realise it, but our body is constantly working. There are so many functions and actions that the human body needs to perform. The heart, for example, needs to keep pumping blood constantly. If it stops beating, we stop breathing. So, our body is the perfect example of a complex machine. There are many different cells and tissues that form the organs in our body. These organs work together to perform different tasks.

When different organs work together, we call this an organ system. These organ systems work together for our body to function well as a whole. Each organ system has a specific role. Do you know how many organ systems there are in the human body? Eleven! One of these is the circulatory system. The heart, blood, and blood vessels are part of this organ system. Read on to find out how our circulatory system works.

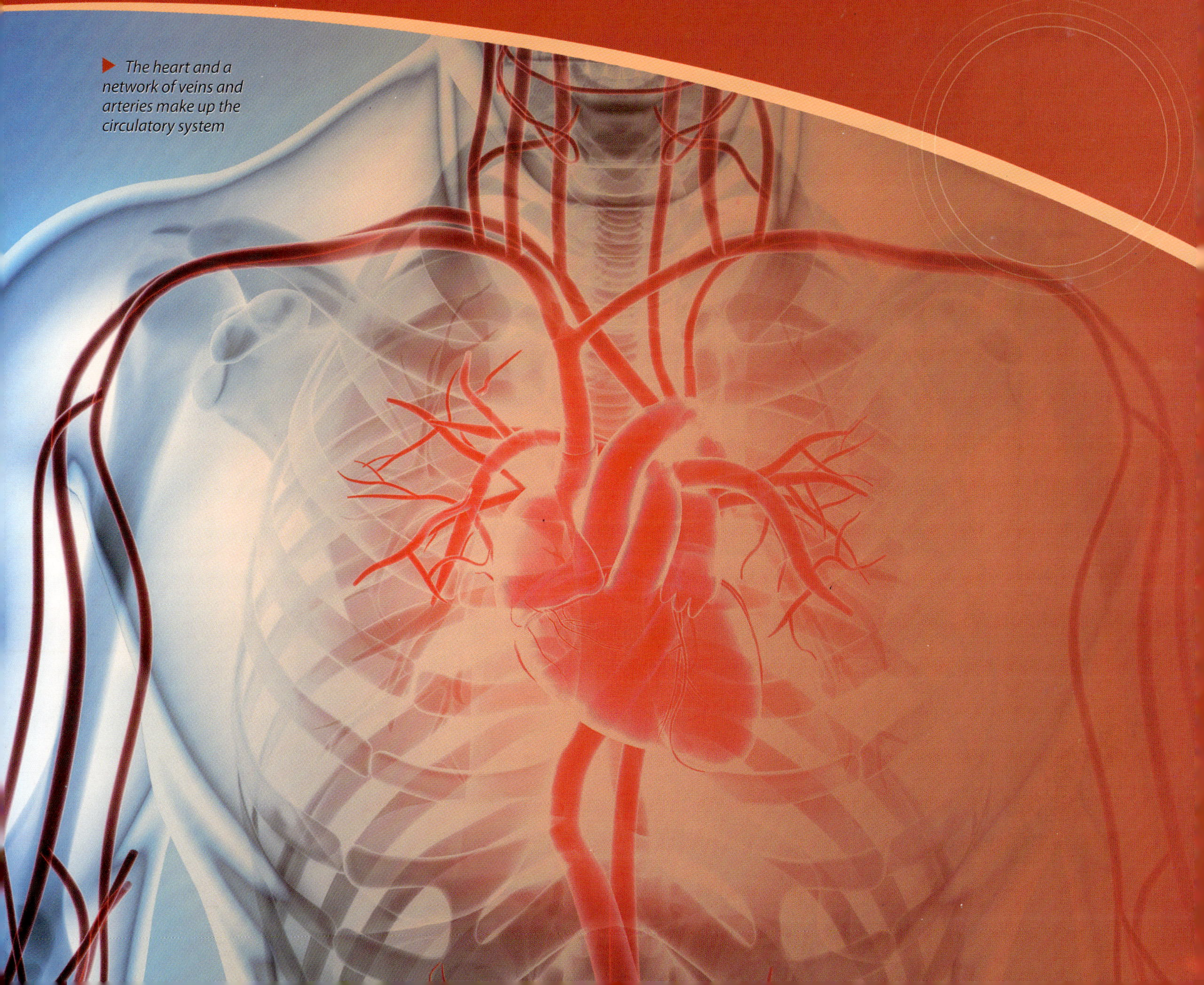

▶ *The heart and a network of veins and arteries make up the circulatory system*

At the Heart of Everything

You might have heard doctors say that the heart is a pump. That is true. This little organ is the size of your fist. It is the second-most important organ in your body after the brain. It pumps blood all the time, to all the organs and tissues of your body, whether you are awake or asleep, running, or sitting. This is how your organs get oxygen and get rid of unwanted carbon dioxide.

▲ Red Blood Cells in a blood vessel

Arteries

Your arteries carry **oxygenated blood** from the heart to all the organs. Once the organs have taken up the oxygen they need, the veins bring the **deoxygenated blood** back to the heart. Only the lungs are an exception, as they get the deoxygenated blood, and give back oxygenated blood to the body.

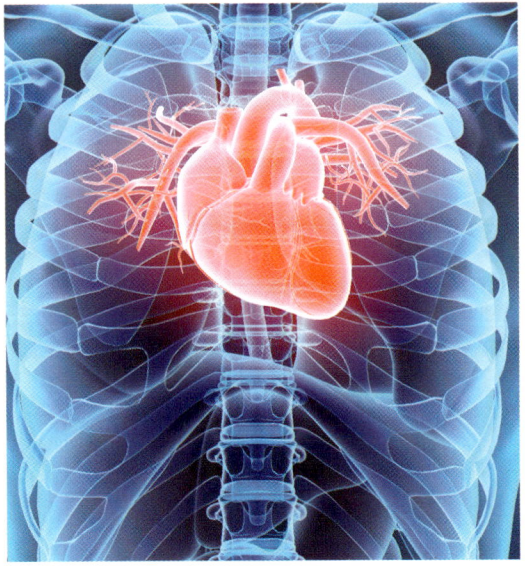

Your Body's Highways

Your heart, **arteries**, and **veins** make up the body's **circulatory system**. Think of it like the system of interstates (USA) or motorways (UK) that connect the whole country. The biggest artery is the aorta, which branches into different arteries for each organ. These, in turn, branch into tiny **capillaries** which go into each tissue of the organ.

Oxygen goes out of the capillaries and into the cells of the organ, while carbon dioxide comes into them. The capillaries join up to form veins, which come out of the organs and join the main vein, the vena cava, which takes blood back to the heart. The whole thing goes on and on, and so your blood 'circulates'.

 ◀ Put your hand over the left side of your chest. Can you feel your heart beating?

Heart Muscles

Your heart is mostly made up of muscles. As the vena cava (see pp.6) pours blood into your heart, the muscles squeeze at one go, pushing the blood out into the aorta. Your muscles require the strength to push blood with so much force that it reaches all the organs, no matter how far away they are from the heart.

The heart keeps pumping all day without rest; if your organs do not get oxygen even for a few minutes, they will collapse and you will die.

In Real Life

Did you know that your heart beats 80 times a minute? It beats faster when you are excited or exercising.

▶ *Wearable pulse meters have become a go-to for everyone nowadays*

Keeping the Heart Safe

As our heart is a vital organ, the body protects it in many ways. In front of the heart is a tough bone called the sternum or breastbone. The breastbone connects to your spine by a number of bones called ribs, which make up the ribcage. The ribcage keeps your heart safe, and protects your lungs.

How Your Heart is Made

Your heart is made up of 'cardiac tissue', which is laid out in three layers. The outermost layer is known as the **pericardium**, which is filled with a liquid that stops the outer wall of the heart from getting dry, so that it does not rub against the lungs. Pericardium also keeps the heart muscles well lubricated. It stops germs from attacking the heart and helps it heal quickly if they do attack it. It makes sure that the heart does not expand too much when being filled with blood, so that your **blood pressure** can be maintained.

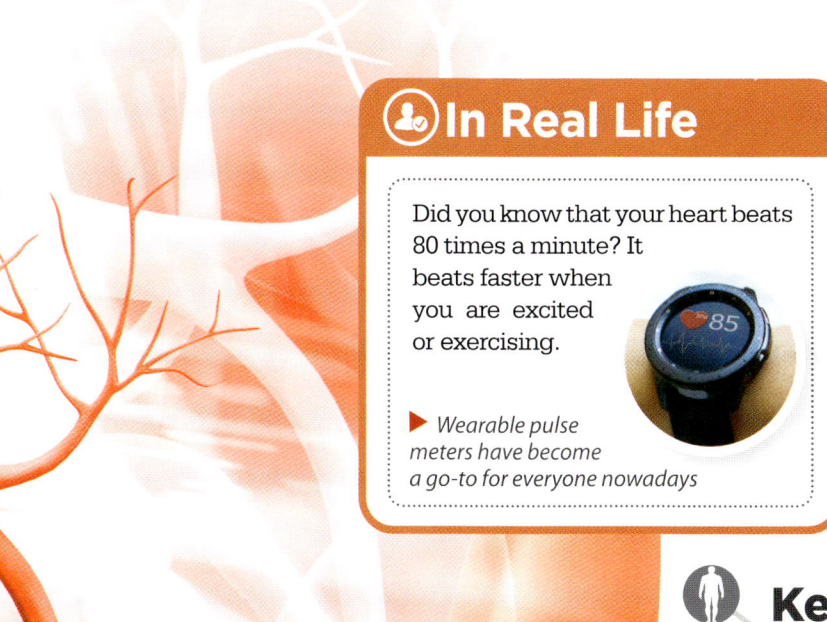

▲ *Your heart pumps about 9,092 litres of blood each day*

◀ *The pericardium anchors your heart to your chest*

▲ *Layers of the heart*

Inside it is the myocardium, which is made of special muscles called cardiac muscles that never get tired. When it contracts, blood is pushed out, and when it expands, new blood fills in. It is the thickest layer. The innermost layer is the endocardium. It is a very thin layer that shields the heart's valves.

Inside Your Heart

Let us try and understand why the heart is so important. When a heart surgeon opens the heart, they see four chambers, separated by the inner heart wall called septum and valves. There are two on the upper, broad part called the base. These chambers are the **atria** (singular: atrium) or auricles, one to the left and one to the right. The other two chambers are the left and right **ventricles**, which are in the apex, the narrower, lower part of the heart.

There are valves between the chambers, just like there are doors between the rooms of your house. They work like gates, stopping blood from leaking out of a heart chamber once it has been filled. This makes the heart's work smooth, and also maintains your blood pressure.

Isn't It Amazing!

The left ventricle is the strongest and largest chamber of the heart, sending blood to all parts of the body. It pumps five times as much blood as the right ventricle.

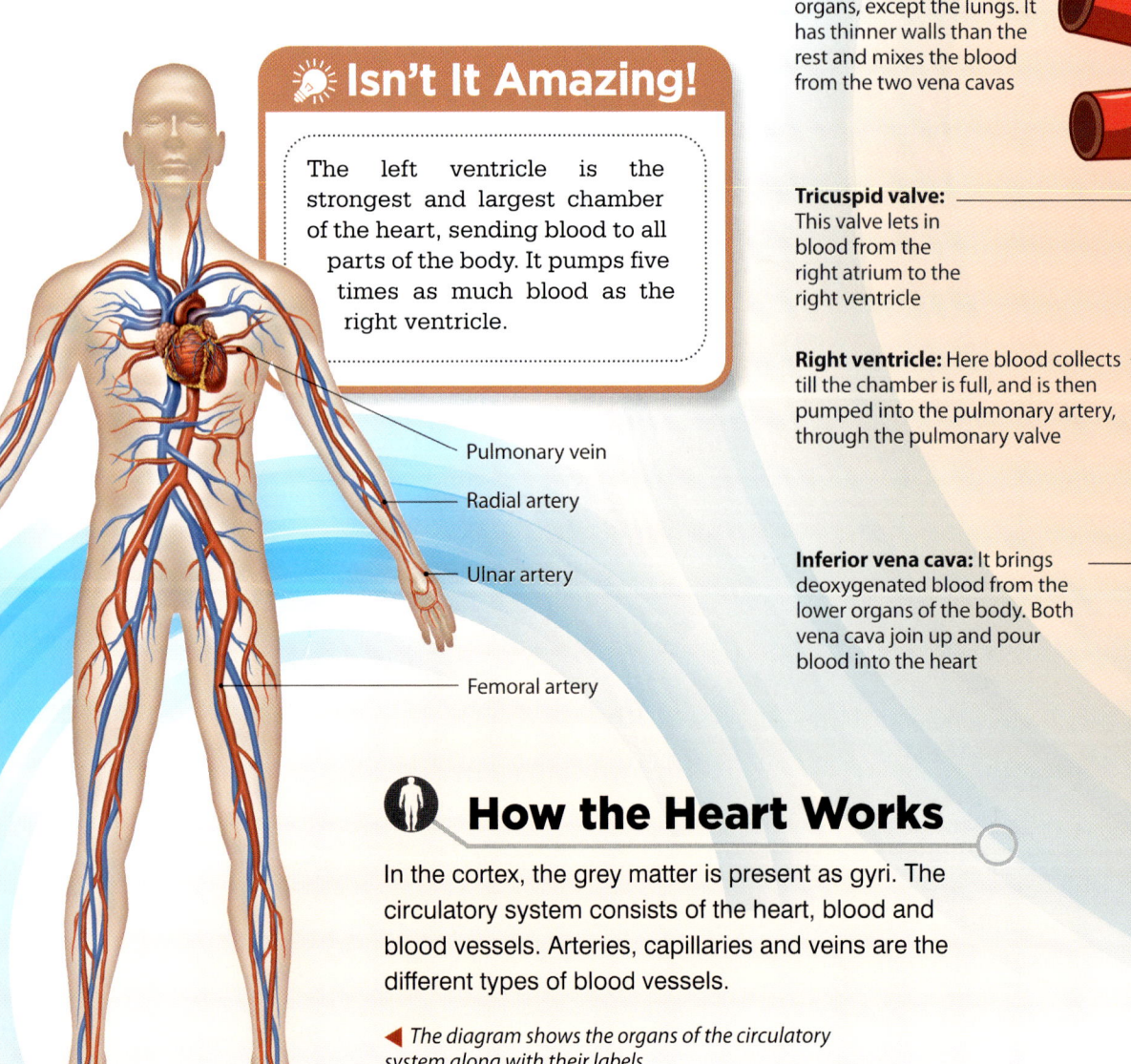

- Pulmonary vein
- Radial artery
- Ulnar artery
- Femoral artery

Superior vena cava: It brings deoxygenated blood from the head and upper organs

Right atrium: This is the first chamber of the heart to receive blood from all organs, except the lungs. It has thinner walls than the rest and mixes the blood from the two vena cavas

Tricuspid valve: This valve lets in blood from the right atrium to the right ventricle

Right ventricle: Here blood collects till the chamber is full, and is then pumped into the pulmonary artery, through the pulmonary valve

Inferior vena cava: It brings deoxygenated blood from the lower organs of the body. Both vena cava join up and pour blood into the heart

How the Heart Works

In the cortex, the grey matter is present as gyri. The circulatory system consists of the heart, blood and blood vessels. Arteries, capillaries and veins are the different types of blood vessels.

◀ *The diagram shows the organs of the circulatory system along with their labels*

Circulatory Cycle

Every drop of blood finishes one cycle when it goes out of the heart carrying oxygen, through the organs, back to the heart, into the lungs and back from the lungs to the heart, where it receives fresh oxygen. This cycle is kept going by the cardiac cycle.

Aorta: This is the body's biggest artery and takes blood to all the organs. It loops around the heart and breaks into two—one going to the head and one to the lower body

Pulmonary artery: This comes out of the heart and branches into two. It carries deoxygenated blood to the left and right lungs, where the carbon dioxide is released into air and blood becomes oxygenated

Pulmonary veins: These bring back blood, freshly oxygenated from the lungs to the left atrium

Left atrium: This has a thicker wall than the right atrium. It collects blood coming from the lungs

Mitral valve: This valve opens when the left atrium is full and lets blood into the left ventricle

Left ventricle: This pumps out the oxygenated blood to the rest of the body through the aorta. It is plugged by the aortic valve

▲ The chambers of the heart are joined by valves

Pumping of the Heart (Cardiac Cycle)

The heart never rests. But it has a quiet diastolic phase when the ventricles are being filled with blood. Once they are full, the heart enters the active systolic phase. The heart muscles contract together, and blood is pumped out with force into the aorta and pulmonary artery. Together these make up one heartbeat. The force with which your heart pumps decides the pressure with which blood flows.

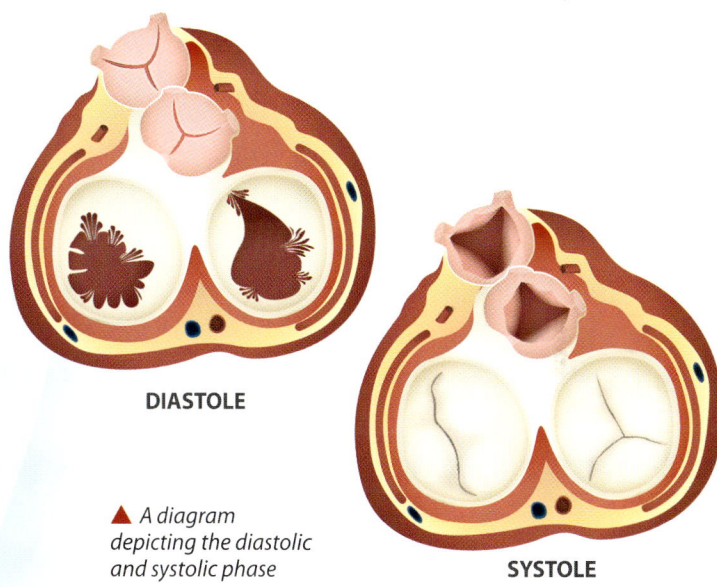

DIASTOLE

SYSTOLE

▲ A diagram depicting the diastolic and systolic phase

Evolution of the Heart

Unlike human beings (who are **vertebrates**), invertebrates have no blood to transport oxygen or nutrients. Air reaches tissues directly through spiracles, while nutrients are absorbed through a fluid called haemolymph. The heart, as an organ, originated in fish—with two chambers: an atrium to receive blood from the gills, and a ventricle to distribute the blood to tissues through arteries. After lungs evolved in amphibians and reptiles, the atrium split into two—one for deoxygenated blood from the tissues, and one for oxygenated blood from the lungs. But the two get mixed in a single ventricle. In crocodiles, birds, and mammals the heart finally becomes four-chambered, and deoxygenated blood is completely separated from oxygenated blood.

Arteries & Veins

Did you know that within a minute, blood from your heart will have reached every cell of your body? That is because our body has an efficient transport system made of arteries and veins, together called blood vessels. Let us take a look at their similarities and differences. The arteries and veins are tube-like, so they ensure that blood flows in a direction. Valves in the arteries and veins make sure that blood does not flow backwards. Blood flows in the arteries away from the heart towards the organs. The oxygenated blood in them makes them look red. In the veins, blood flows towards the heart from the organs. The deoxygenated blood in them makes them look blue.

What Do Arteries Do?

Oxygen is very important for your body's cells as they need it for making energy from the food you eat. As they keep using it, they need more. It is the arteries that bring them the oxygen from the heart, which it gets from the lungs. Other than oxygen, your arteries bring hormones from various glands in the body. Hormones help communicate to the cells what to do and when. Arteries also bring nutrients like glucose and amino acids from the intestines, which digest the food you eat. The pulmonary artery is the only one that's different, as it carries deoxygenated blood from the heart to the lungs for oxygenation.

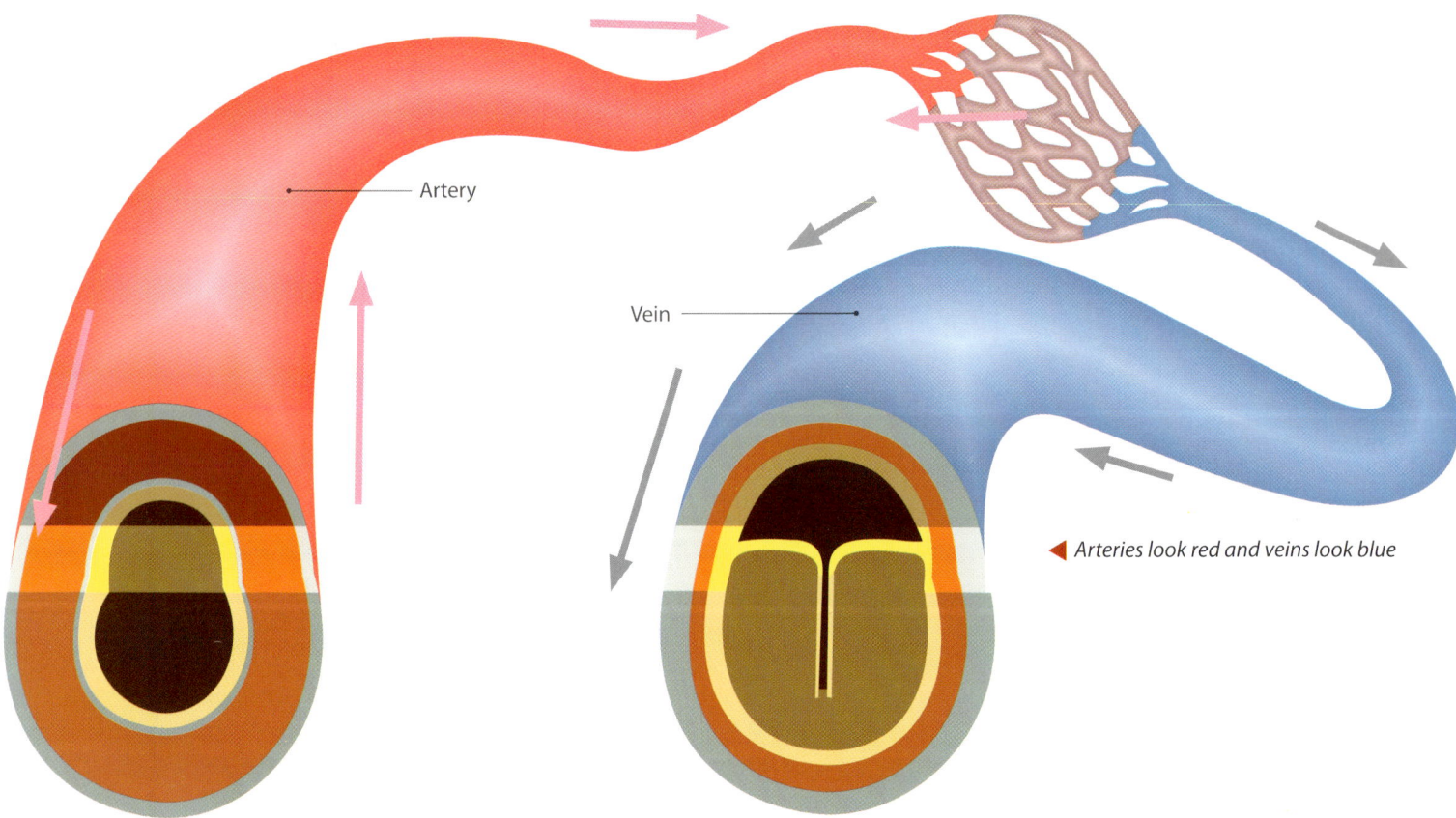

◂ Arteries look red and veins look blue

What Do Veins Do?

Carbon dioxide is what is left once food is turned into energy (or used for making proteins needed by your tissues). It must be removed from the body so that fresh oxygen can be brought in. This job is done by the veins. They also carry away other wastes from the cells, like urea. Two veins are different from the others. The pulmonary veins bring blood rich in oxygen from the lungs to the heart. The hepatic portal vein takes blood with glucose from the intestine to the liver, where the extra glucose is stored.

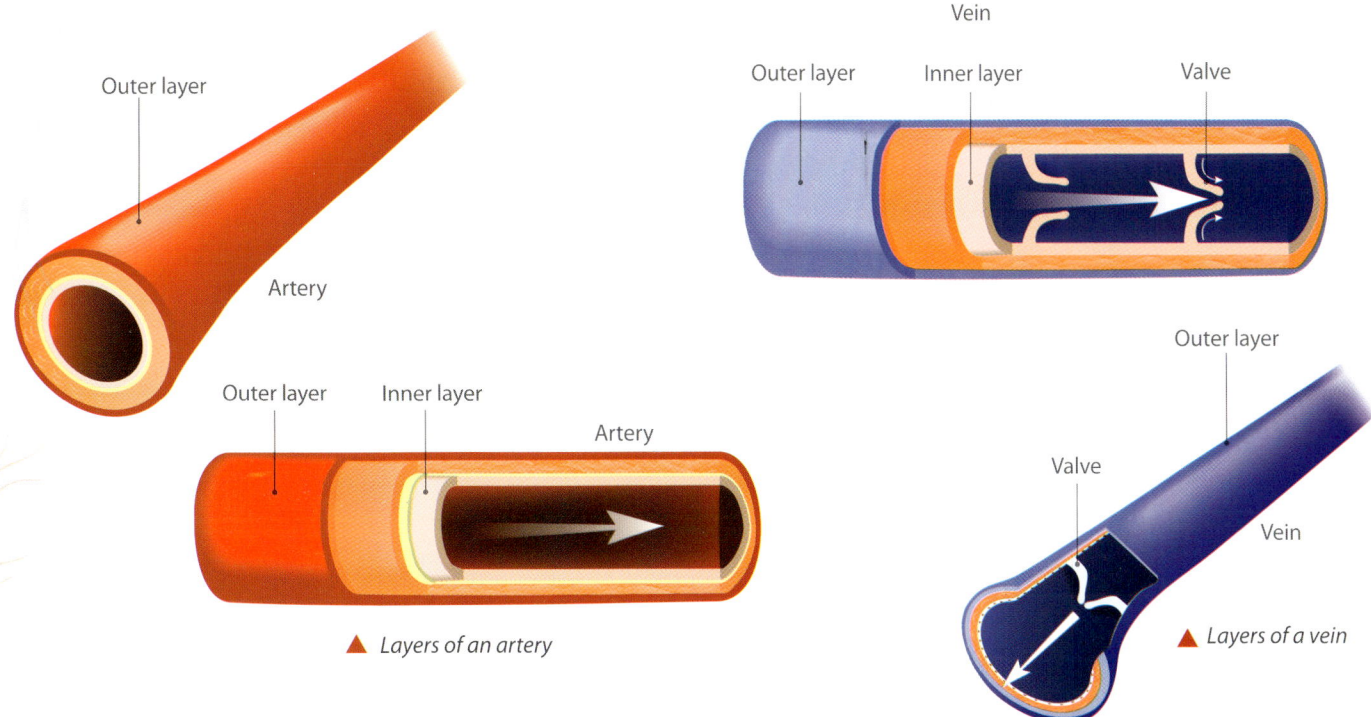

▲ Layers of an artery

▲ Layers of a vein

How are Arteries and Veins Made?

Like your heart, your arteries and veins also have three layers wrapped around each other. The outermost layer is the 'tunica adventitia' which is made of loosely bound cells called 'connective tissue'. Next to it is the 'tunica media', which has tiny muscles that help push the blood along. It is thicker in arteries and thinner in the veins. Innermost is the 'tunica intima', made of a wall of cells that are tightly bound to ensure no leaks occur.

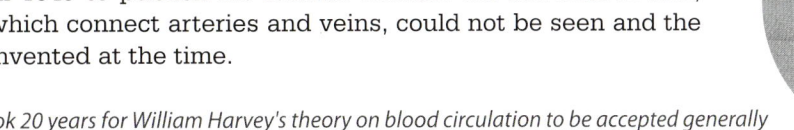 Incredible Individuals

William Harvey (1578–1657) was the personal doctor to King James I of England. He demonstrated that the arterial and venal blood formed a single system and that the heart was the pump that kept blood flowing. Though he discovered circulation in 1618, he waited till 1649 to publish his results. Doctors did not believe him, because capillaries, which connect arteries and veins, could not be seen and the microscope was not invented at the time.

▶ It took 20 years for William Harvey's theory on blood circulation to be accepted generally

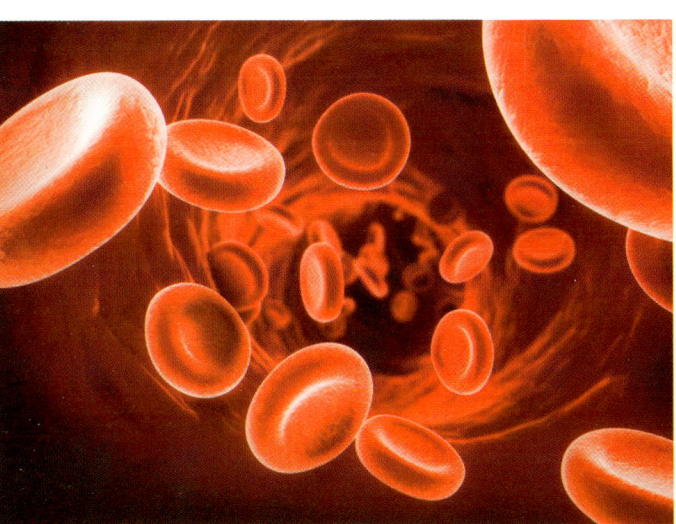

Deep & Superficial Veins

The inferior vena cava is so called that because it lies posterior to the heart. It is the largest vein in the body, collecting deoxygenated blood from all the organs. The veins that bring blood to it are called deep veins, because they are deep inside your body. The veins on your skin are called superficial veins. You can see one stick out of the insides of your elbows. These veins connect to the deep veins.

◀ Red Blood cells are responsible for the transportation of oxygenated and deoxygenated blood

Feeding the Organs

Your aorta starts from the left ventricle of the heart, and gives out branches or arteries that go to each organ. Inside the organ, they branch into smaller arterioles. As they go deeper into the organ, they branch out into really thin blood vessels called capillaries. Did you know that the heart has its own artery that supplies blood to each of the heart's muscles?

Look at one of your hair strands. Some blood vessels are extremely small; that the smallest blood vessels measure 5 mm, that is less than one-third of a hair strand!

▲ Human hair under a microscope

Capillaries

Capillaries act like both arteries and veins. They do not just supply oxygen and nutrients, but also pick up waste and carbon dioxide from the organs including the heart. They join to form venules, which come out of the organ to form the main vein, which in turn joins the vena cava. You can see a network of capillaries if you look at your eye in a mirror by gently pulling down an eyelid.

Precapillaries are located between the capillaries and the smallest arterioles or arteries. These are considered to be intermediate vessels. Precapillaries control how the capillaries are emptied and filled. They have muscle fibres, unlike the capillaries.

Types of Capillary

Your body has capillaries of three types. They are the continuous capillary, the fenestrated capillary, and the sinusoid or discontinuous capillary.

Continuous Capillary

These capillaries have thin gaps between the cells that make up its walls. This lets the fluid part of blood go out into the tissue, and come back in. This kind is found in the lungs, muscles, and nervous system.

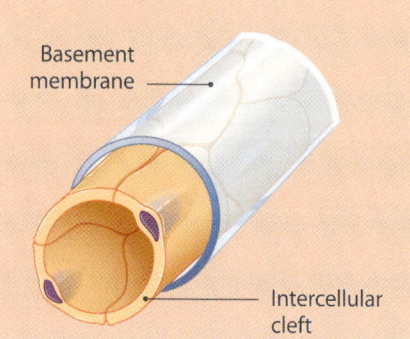

Fenestrated Capillary

The fenestrated capillaries have little holes (like a shower head) that let blood go out and come in. This kind runs in the kidneys, intestines, and glands.

▼ Capillaries are usually 10 μm in size

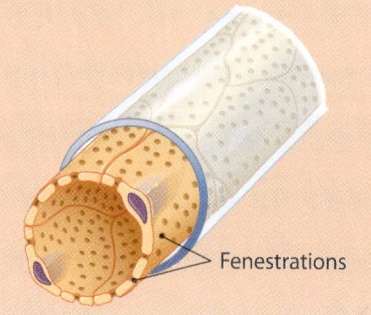

Sinusoid Capillary

Lastly, the sinusoid or discontinuous capillaries have really big gaps that allow blood cells to get into the tissue. You see them in the liver; spleen, where damaged blood cells are destroyed; and bone marrow, where new blood cells are born.

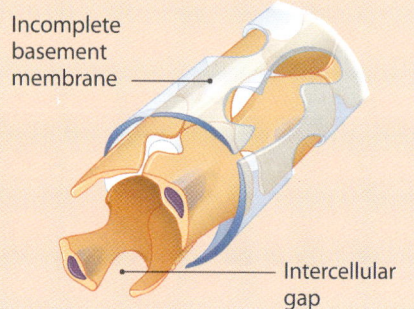

Keeping the Beat

As the heart beats, it pushes blood into the arteries. The blood pushed in with each beat is called a pulse. The number of pulses every minute is therefore called your 'pulse rate'. Doctors measure your pulse rate when you go to them to find out what is making you ill and whether the cause of your illness is affecting your heart. Patients with arrhythmia have an irregular pulse and may need a pacemaker.

Know Your Pulse Rate

Take the index and middle fingers of one hand and press them softly on the thumb-side of your wrist on the other hand. You can feel the pulsations. Look at a clock and count how many beats you feel every ten seconds. Multiply by six—that is your pulse rate.

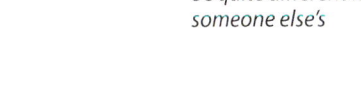

▲ Your pulse rate can be quite different from someone else's

Incredible Individuals

Did you know that Wilson Greatbatch (1919–2011), who invented the pacemaker, was not a doctor, but an electrical engineer? His device has saved millions of lives.

▲ Greatbatch had over 325 patents to his name!

Keeping the Rhythm

The beating of the heart is controlled by two nerves. The accelerans nerve speeds up the heartbeat and the vagus nerve slows it down. Arrhythmias occur when one or both of these nerves do not work properly, or when the muscles of the heart do not correctly respond to these nerves.

A normal pulse rate is an indicator of good health, and hence it needs to be maintained properly. When this rhythm is adversely affected, doctors try to fix it using a **pacemaker**. Artificial pacemakers help correct signals to the heart muscles by giving them tiny electric pulses.

Patients who wear a pacemaker need to be careful. Doctors usually ask them to avoid stressful activities, like lifting heavy loads or climbing many stairs. Patients must also keep away from anything that may upset the pacemaker's generators, like cell phones, walk-through metal detectors and medical equipment like MRI scanners.

▼ A pacemaker is a small device, about the size of a matchbox or smaller

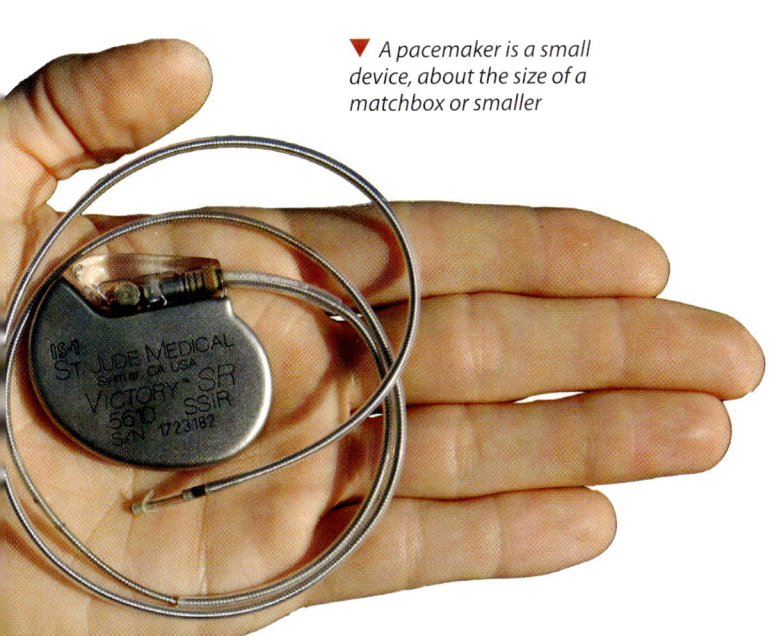

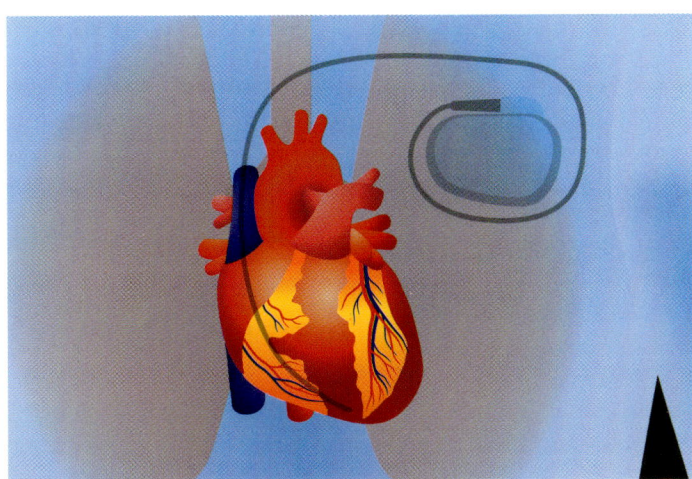

▲ A pacemaker is surgically implanted in your chest

Heart Troubles

By now, we know that the heart plays the most important role in the circulatory system. So, what happens when it takes ill? Like any other organ, the heart too has its share of diseases and disorders. But unlike other organs, diseases of the heart often cause death, because when it stops working, other organs are soon starved of blood and oxygen, they in turn, stop functioning. Let us try and understand the major problems the heart can have. Some are common and can affect anybody at any time, while others are rarer and affect only some kinds of people.

Heart Attack

A heart attack happens when the muscles of the heart do not get enough oxygen-rich blood. A blood clot in the artery of the heart reduces the amount of blood flowing and blocks the passage of blood cells. Some of the heart muscles stop getting oxygen and die. This causes paralysis of the heart, as it cannot pump anymore blood. This is called a '**heart attack**'.

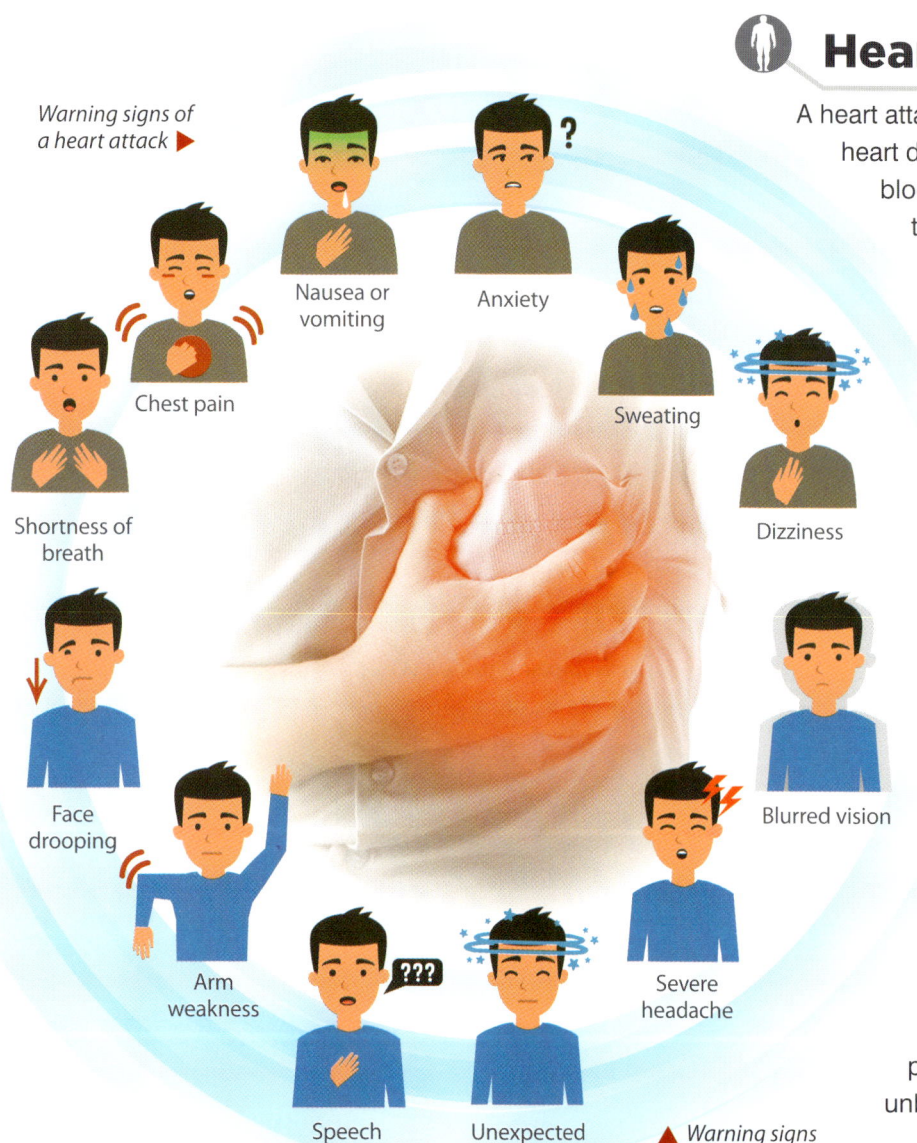

Warning signs of a heart attack ▶

▲ *Warning signs of a stroke*

Heart Failure

When the heart is not able to pump blood properly, it is called heart failure. A heart failure does not mean that the heart stops working completely. Instead, it does not have enough pumping force, and doctors see this as low blood pressure. This also means that your body is not getting all the oxygen it needs. You may also get heart failure if you are very excited or frightened, because the heart is pumping faster than normal, building up unbearable pressure.

Stroke

Blockage of the arteries of other organs can cause them to fail too. If the artery to the brain fails, the person gets a **stroke**. The brain's cells begin to die without oxygen, and the brain stops sending signals to the rest of the body, which means that they begin to die too. One of these signals is to the heart, telling it to keep beating. A patient with a stroke must get medical help immediately, before the heart also stops beating.

In Real Life

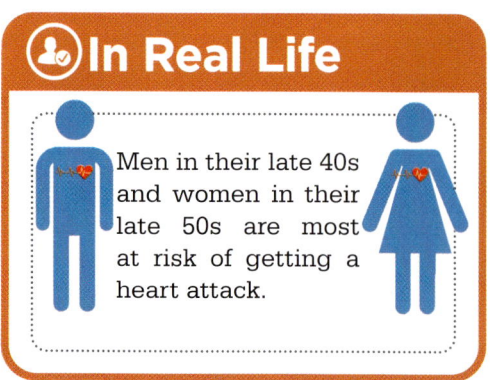

Men in their late 40s and women in their late 50s are most at risk of getting a heart attack.

Aneurysm

As our arteries carry oxygenated blood to our organs, keeping up the right pressure of blood is important. The organs' rate of taking up oxygen depends on that. If an artery swells up, or its walls become weak, then the person can get an aneurysm. If blood pressure rises due to any reason, the aneurysm may cause the artery to burst, which leads to internal bleeding. Blood flows into the patient's body cavity instead of the organs, its pressure drops sharply, and the patient is at death risk.

Sudden Cardiac Arrest

The heart keeps beating as it gets signals from the brain. A problem in the nerves may cause these signals to stop. The heart stops working, and doctors call this 'sudden cardiac arrest'. Patients stop breathing and lose consciousness. It causes instant death if not treated immediately.

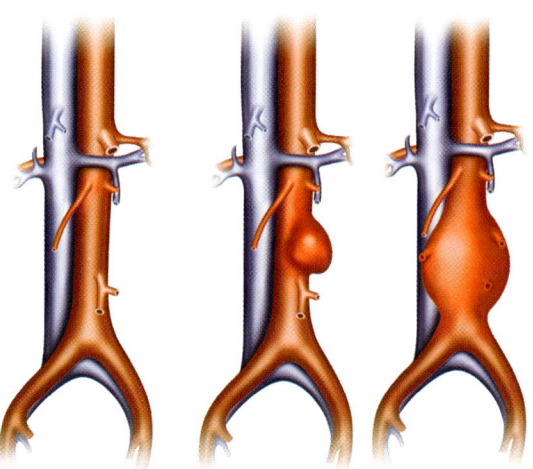
▲ Aneurysm often occur in the aorta, brain, back of the knee, intestine, or spleen

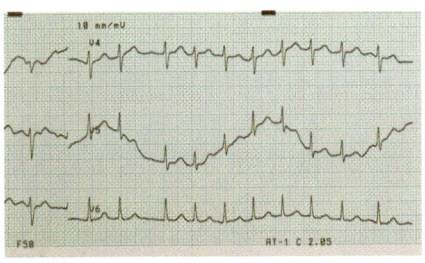
▲ An ECG report shows the electrical activity of your heart at rest

Arrhythmia

Your heart beats at a regular rate or rhythm throughout life. Arrhythmia happens when the heart beats too fast, too slow, or skips beats. This disturbs blood pressure and can cause problems.

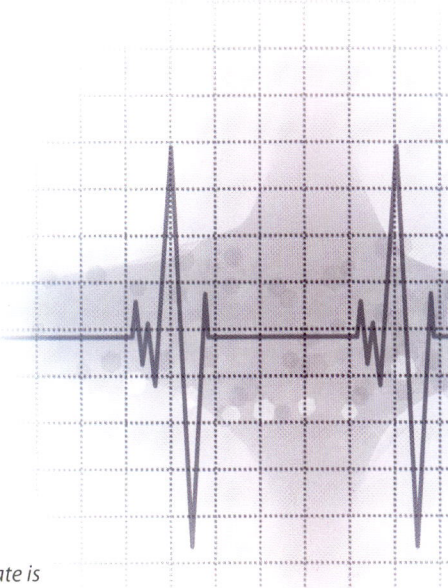
▶ The heart rate is checked and noted on a chart with these kinds of lines called waves

The Beat Regulator

There are many complications that a heart can face. Luckily, with the amazing advancements made in medicine and technology, there is relief even for heart patients, one of them being a heart pacemaker.

The Pacemaker

A pacemaker is a small device which helps to regulate and normalise the rhythm of the heartbeat. Pacemakers are primarily used for patients suffering from arrhythmia. It is placed in the chest, under the skin, with minor surgery.

The size of a pacemaker varies from about the size of a child's palm to as small as a capsule pill. It has two parts: the generator and the leads. The generator has the electrical circuit and the battery for the pacemaker while the leads are a couple of wires that carry an electrical message to the heart. Pacemakers can be placed for temporary or short-term use; or in the case of long-term heart problems, they can be permanent.

A pacemaker works on electrical pulses. It is an electrically charged medical device which prompts the heart to beat at a normal rate.

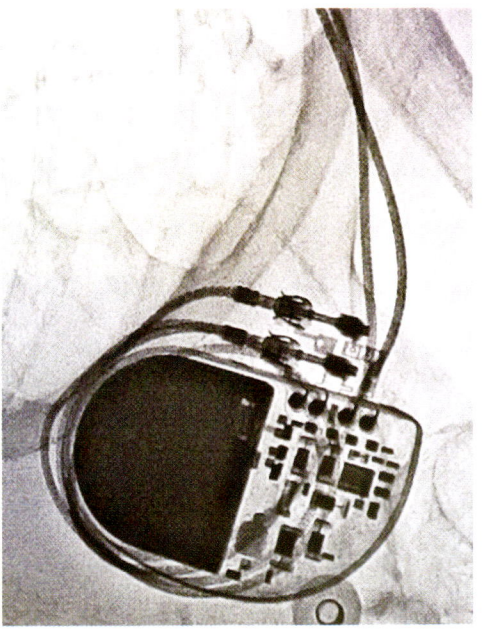
▲ An x-ray showing a pacemaker fitted in a patient's chest

Transplants
Gifting a New Life

People with a weak heart might not live for very long. But if treatment has not worked for them, do they have hope? Yes! They might receive a new heart which was 'donated' by a brain-dead person, i.e., a person whose brain has stopped working, but their heart can be transplanted to someone else. A **heart transplant** is a very difficult surgery. Prior to it, a surgeon checks the donor's compatibility with the patient and if the patient is healthy enough for the heart.

In Real Life

A heart must be transplanted within four hours of the heart being taken from the donor. Hence, once doctors have declared someone 'brain dead', everything must be kept ready for the patient to get the heart. Doctors, police and city authorities cooperate to create 'green corridors' between hospitals, so that an ambulance (sometimes even a plane or helicopter) can transport the heart.

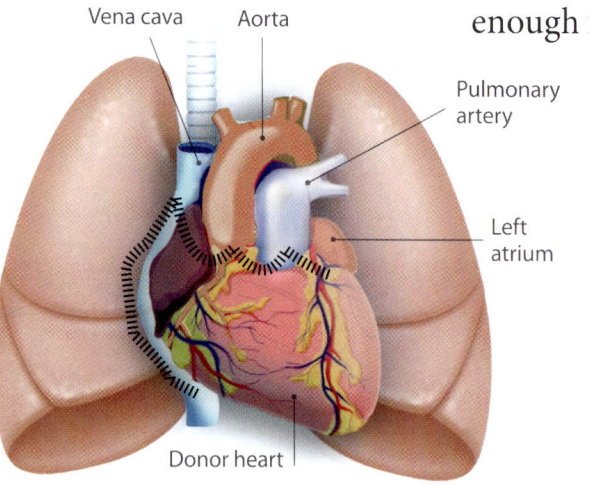

▲ The diagram shows a donor heart after heart transplant surgery

Who Gets a Heart Transplant?

A person gets a heart transplant at a time when their heart is at a very critical stage. It means that none of the medication, treatment, or surgery has worked in curing their heart problems. The heart is donated by a deceased person. This way, the person getting a heart transplant gets a new heart from another human being.

It is not easy to get a donor heart. Thus, it is important to make sure the person receiving the heart transplant really needs it. It is more important to make sure that the patient is healthy enough to survive after getting a new heart. A heart transplant must usually happen within four hours of the organ being removed from the deceased.

Incredible Individuals

Dr Christiaan Barnard (1922–2001) performed the first ever human-to-human heart transplant on 3 December 1967. The entire operation took nine hours and required a 30 person team.

▶ Dr Christiaan Barnard

Eligibility

Not everyone needs a heart transplant. Some that do need it, might still not get one because there are few donor hearts available. Here is how doctors decide on a transplant:

✓ The patient must meet health standards before going into the surgery and be healthy enough to heal after it.

✓ The patient should not have a serious disease like diabetes or cancer that could damage their heart.

✓ They must not have an infection at the time of surgery.

✓ They should be ready to change their lifestyle to keep their new heart healthy.

▼ During a heart transplant surgery, the patient is placed on a heart-lung machine

Know Your Blood

What is the composition of blood and why is it so important for our life? It is made of two parts—cells and plasma. Cells are born in the bone marrow and come in three types—red blood cells (RBCs), white blood cells (WBCs), and platelets. Plasma is made of water, salts, hormones, and many other biochemicals.

Blood makes up 7–8 per cent of your body's weight. Depending on your age, size, and gender, you may have between 4–6 liter of it in your body.

The cornea of the eye can get oxygen directly from the air. All other organs of your body must get their oxygen from the blood.

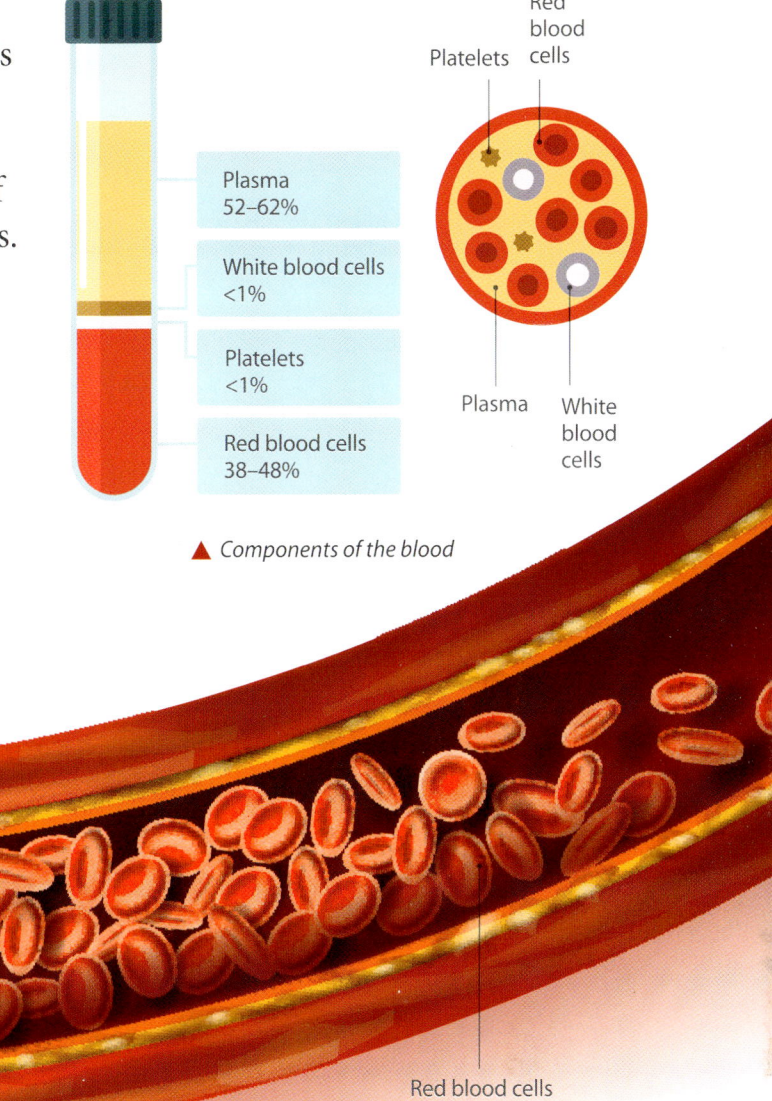

▲ Components of the blood

▲ Nearly 7 per cent of your body weight is blood

 ## Why is Blood Important?

The primary role of blood is to carry oxygen and nutrients to the cells of your body, and to carry out waste products to the lungs and kidneys. It heals wounds, by forming clots, it takes poisons to the liver for detoxification; and also helps your body maintain its temperature.

 ## Homeostasis

This is the word used by doctors and scientists to describe how your body maintains its temperature. When the weather is cold, the body can lose heat. The brain makes the superficial veins and arteries contract, so that less blood flows through them, and less heat is lost through the skin. In hot weather, the opposite happens.
Blood vessels expand, more blood flows and more heat escapes through the skin.

Blood Cells
Little Transporters & Defenders

The cells of your blood do most of its important jobs. The most in number are the red blood cells (RBCs) or erythrocytes. They give your blood its colour and carry oxygen from the lungs to the organs through the heart. Each drop of blood has about 5 million RBCs. On the other hand, white blood cells (WBCs) or leucocytes make up less than 1 per cent of blood. They help defend the body by swallowing any germs that may enter your body. Both RBCs and WBCs are made in the bone marrow, and at the end of their lives, are destroyed in the spleen and liver.

▲ *You can donate RBCs to hospital blood banks. They specially freeze them, till somebody needs a blood transfusion. The RBCs can be stored for 42 days*

Isn't It Amazing!

If there are 5 million RBCs in one drop of blood, how many are there in the whole body? 30 trillion! Every fourth cell in your body is an RBC.

Red Blood Cells

Under a microscope, RBCs look like disc-shaped doughnuts (without the hole), making up 40 per cent of your blood. They travel through your body and live for about four months. Each second, 2 million new RBCs are born, and 2 million die. They are made from 'haematopoietic stem cells' in the bone marrow; it takes about a week for them to turn into 'mature' RBCs. Then they enter the blood through the 'sinusoid capillaries' to get on with their jobs!

What Do Red Blood Cells Do?

RBCs have a special protein in them called **haemoglobin**, which contains iron. The iron makes haemoglobin, RBCs and therefore your blood red in colour. In the lungs, each unit of haemoglobin binds four molecules of oxygen and carries them to the rest of the body's tissues, which are hungrily waiting for them. They also take carbon dioxide from the tissues to the lungs, where it is released into the air.

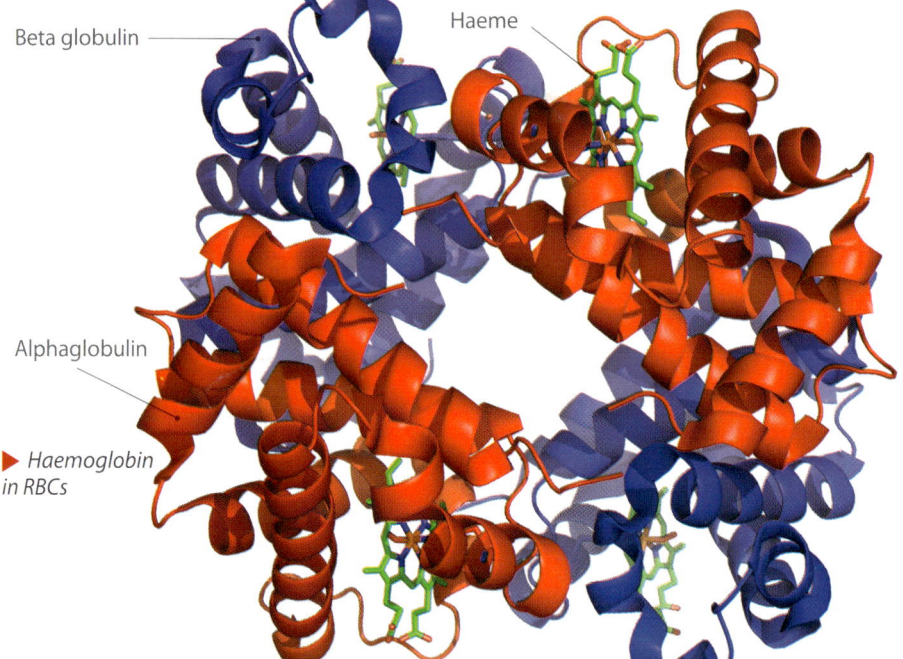

▶ *Haemoglobin in RBCs* — Beta globulin, Haeme, Alphaglobulin

HUMAN BODY — HEART & CIRCULATORY SYSTEM

White Blood Cells

WBCs come in many types—**macrophages**, lymphocytes, neutrophils, basophils, and eosinophils. Each type of cell does something different in protecting the body from germs and allergens. They are actually a part of the immune system, where the body fights diseases and infections. Some WBCs live only for a day, while some may live for several years, remembering a previous infection just in case you are infected again. WBCs cannot be donated easily as it is hard to extract them from blood.

How Do White Blood Cells Work?

The different types of WBCs work together like a team to fight germs. Along with the lymphatic system, they make up the immune system. Here is what each WBC does.

▲ White Blood Cell (WBC)

Types of White Blood Cell

Neutrophil — Neutrophils go first. They attack most of the harmful germs and send signals to other WBCs to help.

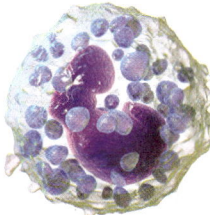

Basophil — Basophils cause inflammation if an allergy-causing substance such as pollen or nuts enters the body.

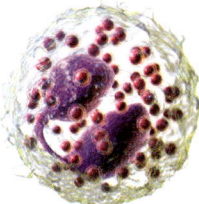

Eosinophil — Eosinophils kill bacteria and parasites in our body and clean up dead cells.

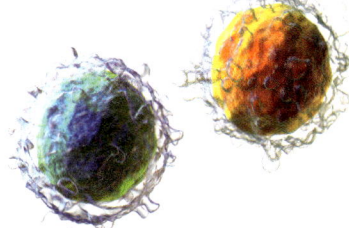

Lymphocyte — Lymphocytes come in many types themselves, making antibodies and other biochemical weapons.

Monocyte — Monocytes transform into macrophages, which swallow infected cells, germs, and dead cells.

Isn't It Amazing!

The cells also interact with each other through special biochemicals called cytokines and interleukins.

Platelets & Plasma

Platelet and plasma cells make up the rest of your blood. Plasma is the liquid part of blood and makes up more than half the volume of blood. It carries the nutrients your cells need, like the salts, glucose, amino acids, and lipoproteins dissolved in it. It is necessary for blood to flow smoothly, with an optimum blood pressure. On the other hand, platelets are the body's tiniest cells and make up a tiny fraction of blood but do important jobs like clotting of blood.

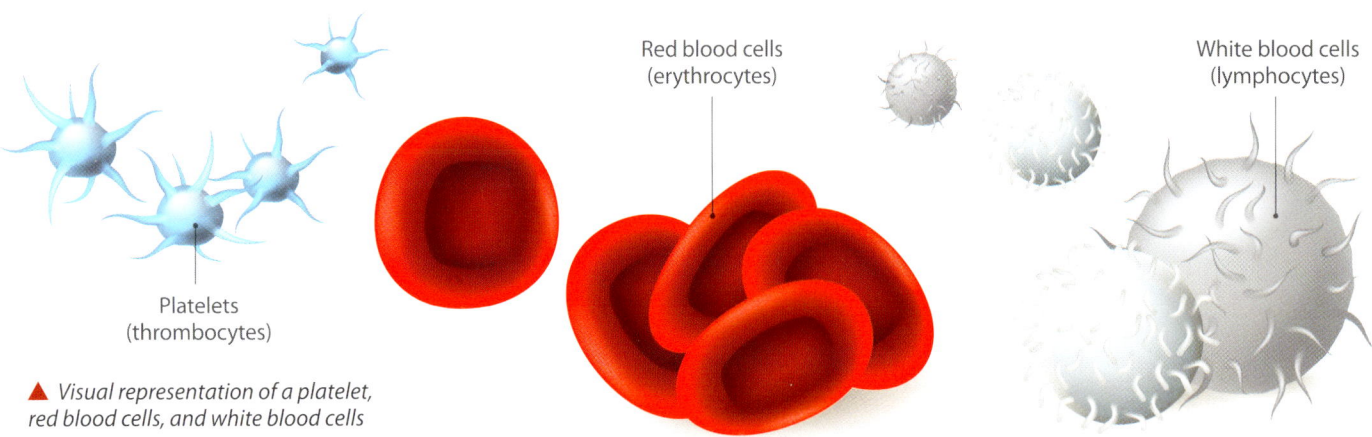

▲ *Visual representation of a platelet, red blood cells, and white blood cells*

Platelets

Also called thrombocytes, these flat cells look like small plates under a microscope. When you get injured, platelets rush to the wound and clog it. Because they are sticky, they make a self-sealing bandage and do not let germs in. This is called clotting and stops you from bleeding. Without platelets, your body would be drained out of blood.

Unfortunately, you can only have a limited number of them. If your platelet count exceeds the limit, they can cause internal clots leading to aneurysms, heart attacks, and strokes.

Platelets can be taken out of the blood and donated separately. This is useful for people with dengue and other diseases where their natural platelet count drops sharply.

Plasma

Plasma is 90 per cent water, so it is natural that it is the main tool of the body to maintain its electrolyte and fluid balance, called 'osmolarity'. Most people suffering from blood loss really just need plasma and/or platelets. The plasma is taken out and stored in blood bags. Unlike cells, plasma can be frozen and stored for a year.

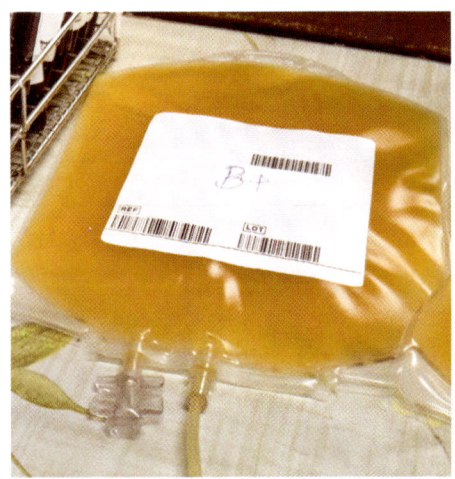

▲ *Frozen blood plasma*

In Real Life

In a plasma donation centre, the blood you give is put into a machine called a centrifuge, which spins it very fast. The cells settle down at the bottom, and the yellowish plasma floats above.

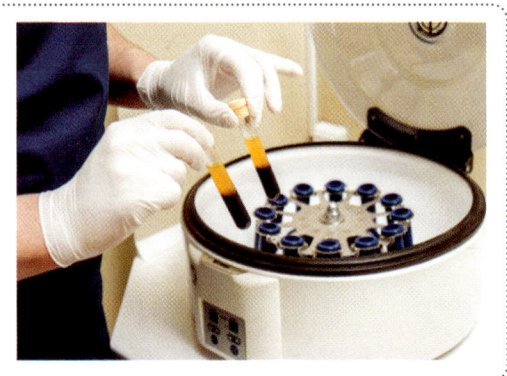

▶ *Centrifuge uses centrifugal force to separate the blood components*

Donate Blood, Save a Life

Blood loss often happens if you have had a terrible accident or are undergoing major surgery. This can lead to oxygen starvation, as there is not enough blood to take oxygen from the lungs to the tissues. This, in turn, can cause organ failure and death.

Patients who need blood depend on people who donate it voluntarily (donors). When blood is injected into a patient, it is called a **blood transfusion**. In modern times, one does not always require all components of the blood to be transfused. Instead, doctors often call for individual blood components, usually plasma, platelets, or RBCs.

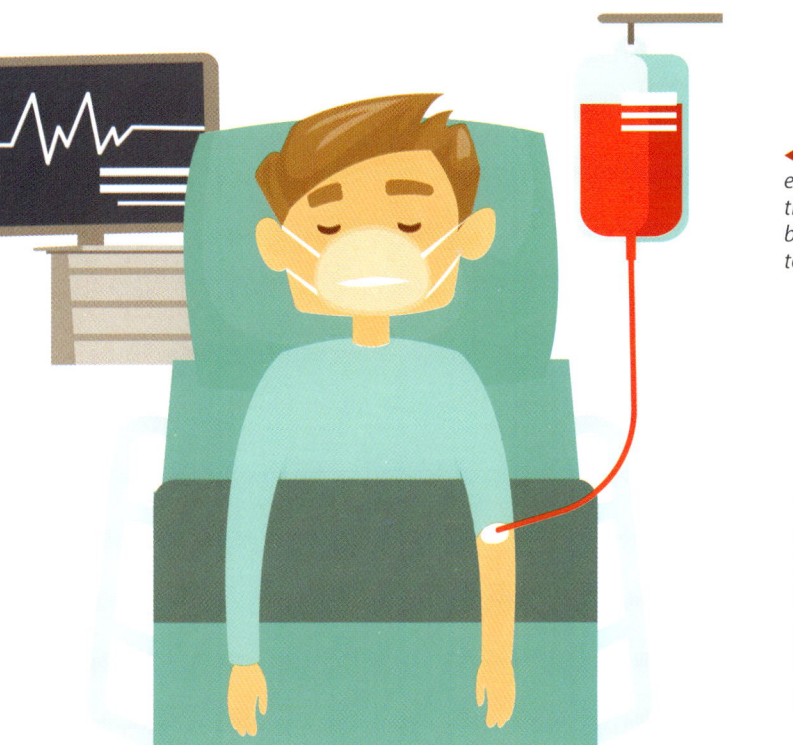

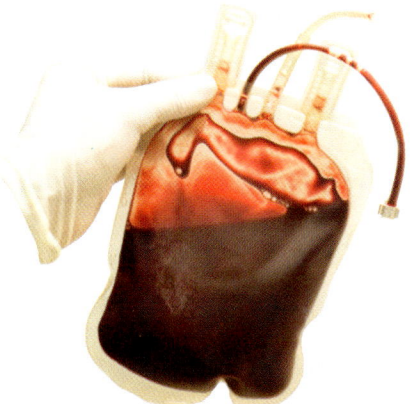

◄ It is not always easy to get a blood transfusion, but blood banks go a long way to help

In Real Life

World Blood Donor Day falls on 14 June around the world. Celebrate it by donating blood at your nearest hospital.

Who can Donate Blood?

To donate blood, you have to be healthy and grown-up, between the age of 18 and 60 years old, and not weigh less than 45 kilograms (99 pounds). If you are anaemic or suffer from an infectious disease, or if you had hepatitis in the past year, you cannot donate blood.

How to Donate Blood?

A grown-up can donate up to 350 mililiters at a time, once every three months. Doctors will place a needle in the vein at the elbow and draw out blood into a specially made blood bag which stops it from clotting or getting infected. Some of the blood is used for testing and the rest is frozen till someone needs it. After a donation, you may feel dizzy or weak because of the loss of fluid. Drink some juice and eat something, and you will be back to normal in a few hours. Your body will replace the plasma within 48 hours and all the RBCs in 4–6 weeks.

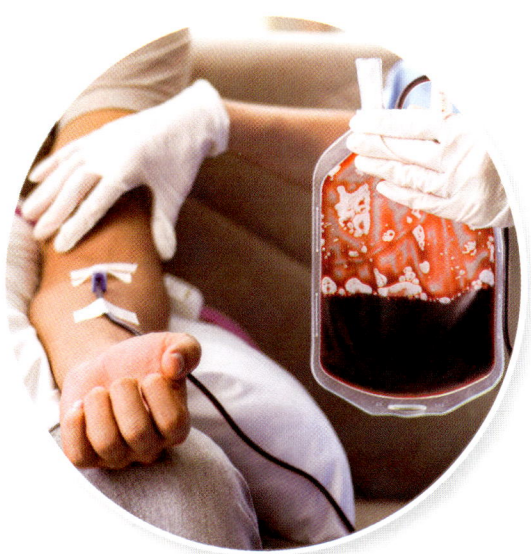

▲ Many neighbourhoods conduct blood drives where healthy people donate blood to local hospitals

Know Your Blood Type

There are two ways in which everybody's blood differs. The first is the ABO system, because of which we have A, B, AB, and O blood types. The other is the rhesus factor, usually written as Rh positive (+) or Rh negative (−). Thus, you get eight types: A+, A−, B+, B−, AB+, AB−, O+, and O−.

Your immune system will only accept blood that matches the type that runs in your circulatory system and reject everything else. Therefore, you need to know your own blood type before you donate blood or receive any. Blood transfusion with mismatched blood leads to medical complications and sometimes even death.

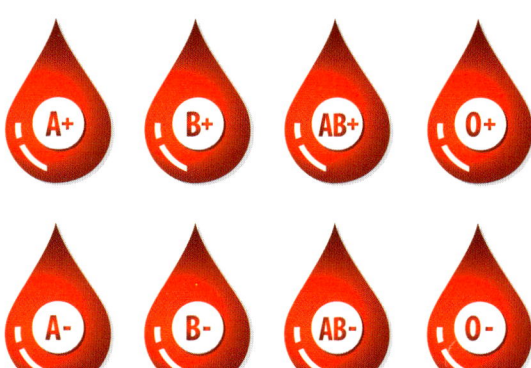

▲ The eight common blood types

Why Must Blood Type be Matched?

The RBCs of your bodies have different 'antigens' on their surface. These are proteins that tell your WBCs that these cells belong to the same body. Some people have a type called A, so they belong to Blood Group A. Their bodies will make antibodies against the other types. People of Blood Group B have another type called B. Some people have both types that is, Blood Group AB, and some have none that is, Blood Group O. The rhesus factor works in the same way.

If you are of type B+ and receive blood of type A−, it may have antibodies that attack B+ RBCs, and kill them. Your body will develop antibodies against A−. These are called transfusion reactions and may cause severe problems in your body. That is why it is important to match. And that is why schools, universities, and workplaces put your blood group on your identity cards.

Antigens and Antibodies in the Blood

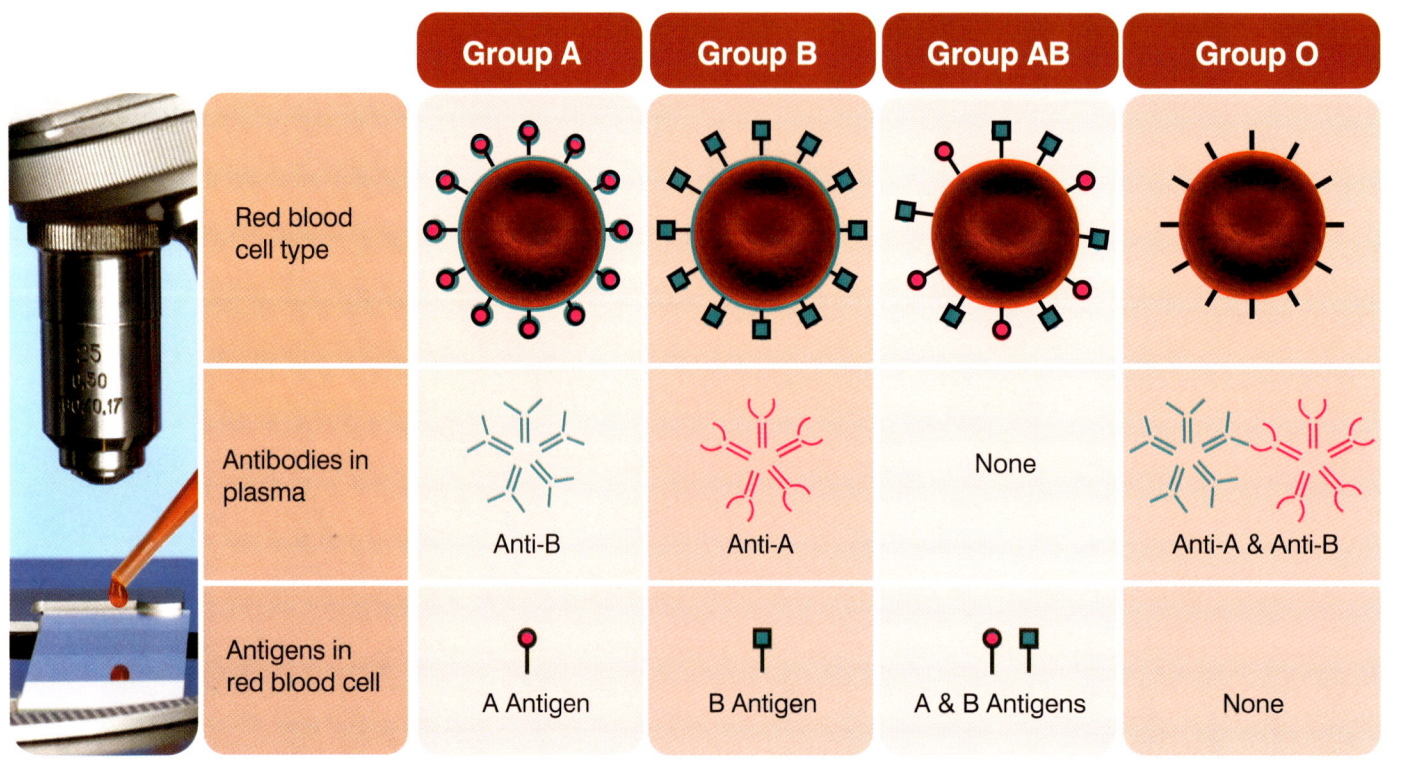

HUMAN BODY | HEART & CIRCULATORY SYSTEM

⭐ Incredible Individuals

Have you heard about the Man with the Golden Arm? His name is James Harrison, a Australian who donated blood for 60 years! Thanks to his donation, 2.4 million children suffering from a disease called Rhesus were saved. This became possible because his blood was found to have rare antibodies that could fight the disease.

▶ James Harrison made his last donation in May 2018, since blood donation from those past the age of 81 is prohibited in Australia

How Does One Match Blood Groups?

If you need blood, doctors will insist on an exact match at the blood bank. But if there isn't any of the right kind, and there is a real emergency, doctors use these rules:

- Give blood of type A to patients who have blood type A or AB.
- Give blood of type B to B or AB patients.
- Give blood of type AB only to AB recipients.
- Give blood of type O to patients of any blood type A, B, AB or O.
- Give Rh positive blood only to those who have Rh positive type, as long as their ABO is matched or follow the above rules.
- Give Rh negative blood to anyone whose ABO matches perfectly or follow the above rules.
- This makes those with O– blood capable of donating blood to anyone. So, people with this blood type are called universal blood donors.
- Those with AB+ type can take blood from anyone, so they are called universal blood recipients.

👤 In Real Life

There are hundreds of rare blood types that people can have because there are more than 600 known antigens. It is the presence or absence of these antigens that determines the rare blood types.

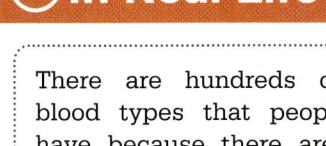

Blood Donation Compatibility Chart

Recipient \ Donor	A+	B+	AB+	O+	A−	B−	AB−	O−
A+	🩸			🩸	🩸			🩸
B+		🩸		🩸		🩸		🩸
AB+	🩸	🩸	🩸	🩸	🩸	🩸	🩸	🩸
O+				🩸				🩸
A−					🩸			🩸
B−						🩸		🩸
AB−					🩸	🩸	🩸	🩸
O−								🩸

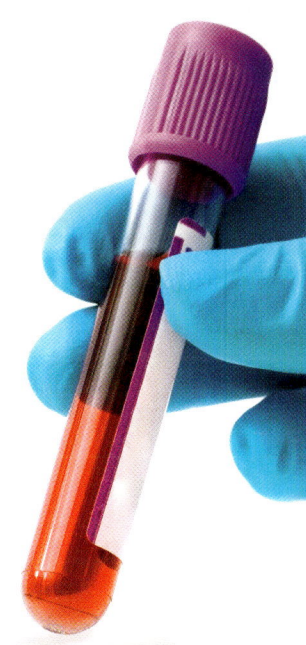

▲ Anti-clotting vials to collect blood

▼ A finger-prick test is done to diagnose diabetes

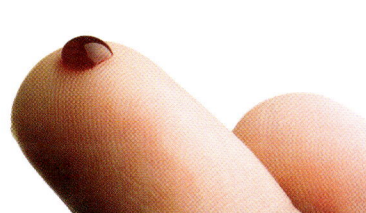

Know Your Blood Pressure

As our heart contracts and pumps blood into our arteries, it puts pressure on the walls of the blood vessels. This is called blood pressure (BP). BP goes up when the heart beats faster and goes down when you are sleeping or doing something quiet, like reading this book. There are actually two pressures, a lower one when the heart is waiting to be filled with blood (diastolic), and a higher one when it pumps out the blood (systolic). One act of pumping and one of relaxing makes up a single pulse.

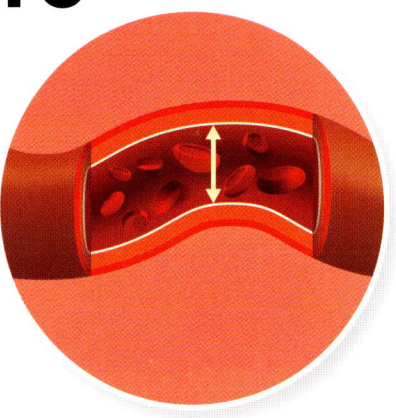

▲ *Blood puts pressure on the walls of the blood vessels*

How Do You Measure BP?

A **sphygmomanometer** measures BP. It has two parts. The doctor will first wrap a rubber cuff around your arm and inflate it to put pressure on the arteries in your arm. The second is the barometer.

Till a few years ago, this was a box that had a tube filled with mercury, like a thermometer. The height of the mercury in the tube (in millimetres/inches) gave the diastolic and systolic pressures. That is why BP is still written down as mmHg, that is millimetres of mercury in the tube. Now, digital sphygmomanometers tell you the BP electronically, along with the pulse. Blood pressure is written as systolic BP and diastolic BP separated by a forward slash, for example 120/80.

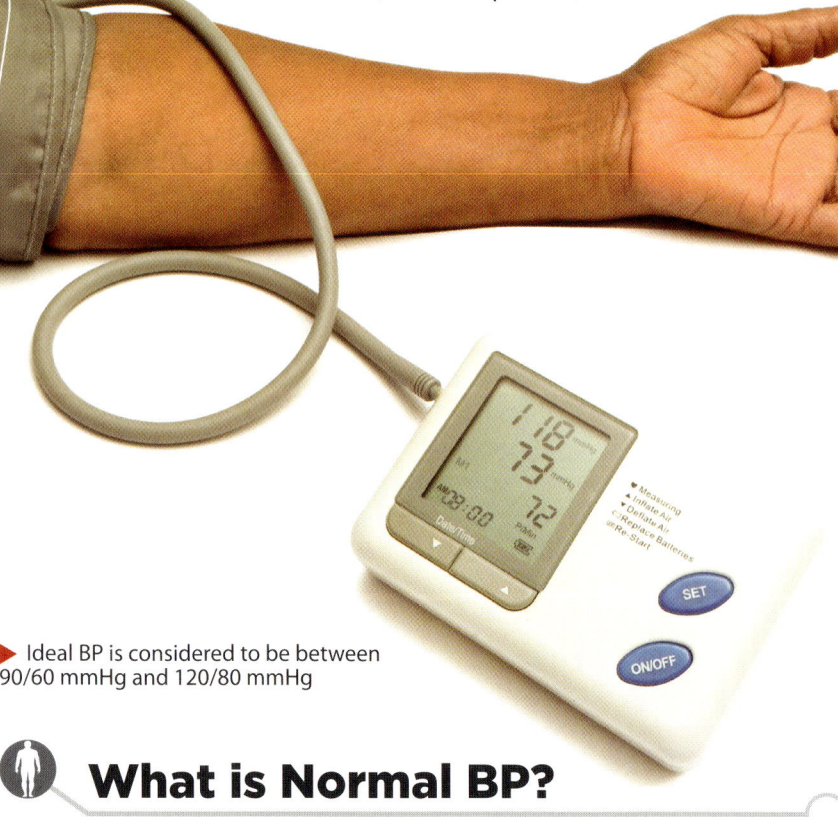

▶ Ideal BP is considered to be between 90/60 mmHg and 120/80 mmHg

What is Normal BP?

Normal BP is actually different for different people, and depends on age, sex, size, and health. But if your BP is regularly above 140/90 mmHg even if you are resting, you have high blood pressure, medically known as **hypertension**. You will feel normal, but you have a higher risk of heart attack, stroke, or kidney failure. On the other hand, if your BP is regularly below 90/60 mmHg, you have low blood pressure, or what doctors call **hypotension**. A sudden drop in BP may make you faint or feel dizzy and light-headed.

In Real Life

Some symptoms for issues with blood pressure are headaches, shortness of breath, nose bleeds, dizziness, nausea, difficulty in concentration, and blurry vision. While some are apparent, most will not show unless it gets serious. Always have regular health check-ups.

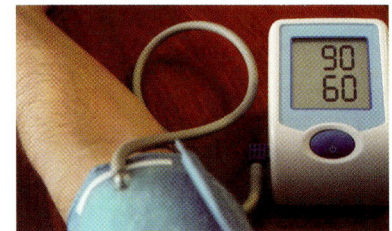

▲ *The reading of 90/60 or lower indicates that the patient has low blood pressure*

Lymph
The Second Circulatory System

Did you know that our body has a second circulatory system, with its own network of organs and vessels? This is the lymphatic system, and it gives the main system a helping hand. It also has WBCs in it, so it also helps the immune system. But more importantly, it helps drain the tissues of waste and excessive fluid, which would otherwise make them swell up and cause a condition called oedema. It recovers plasma that passes through capillaries. The fluid that travels through it is called **lymph**.

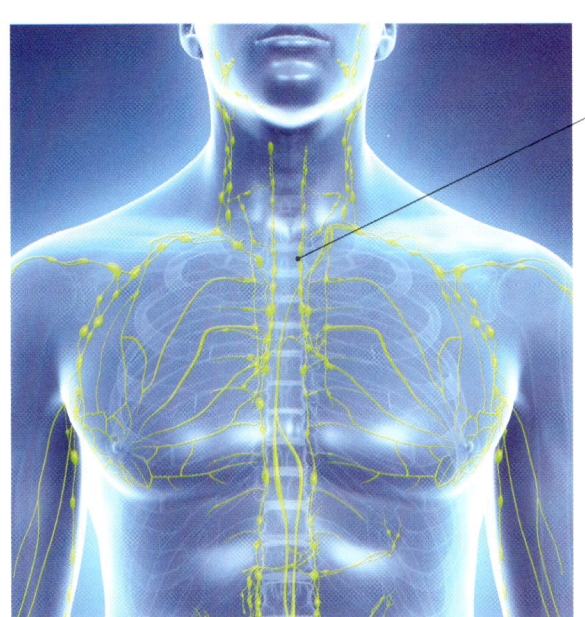

▲ An illustration of the lymphatic system

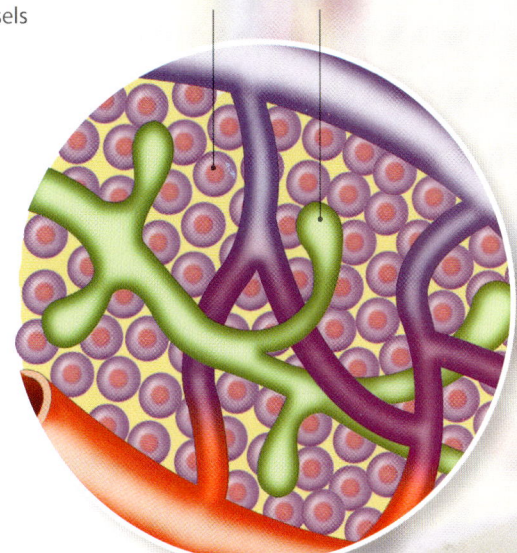

▲ The green lines in the diagram represent the lymph vessels

What is Lymph?

In insects, the lymphatic system is the main circulatory system, as there is a separate respiratory system. But in birds and mammals, blood takes over. But, the plasma from 'fenestrated' and 'sinusoid' capillaries oozes into the tissues, but not all of it comes back. Instead, the fluid becomes lymph, a colourless liquid, thinner than plasma. It flows into lymph vessels, which pump it through lymph nodes and finally pour it into the subclavian vein, near the shoulder bones. Thus, the extra fluid and the nutrients, salts, and hormones dissolved in it slowly come back to the circulatory system.

What Does the Lymphatic System Do?

The lymphatic system does not just carry away the extra fluid. It is also a way for nutrients to reach deep into tissues where blood cannot. From the capillaries, WBCs can crawl into tissues, where they can find and destroy germs. The destroyed germs are carried out through the lymph and finally cleaned up in the liver.

Lymphatic vessels also form part of a special tissue called **MALT**, in the lungs and intestines, where they act as a part of the immune system. Your body has 500 to 600 lymph nodes, which act as resting and training places for WBCs.

Lymphatic System
Your Body's Janitor

The lymphatic system is similar to the circulatory system, but lymph vessels are thinner than blood vessels. They end in two major lymph ducts, the right lymphatic duct and the much larger thoracic duct. Lymph only flows in one direction—back to the heart. For this, it has to travel upwards, as the two lymph ducts meet the circulatory system in the subclavian vein. The ducts have valves to stop lymph from flowing down.

As the lymphatic system does not have its own heart, it does not experience any pressure. Therefore, lymph flows slower than blood. This gives the WBCs in it more time to look for germs and fight them.

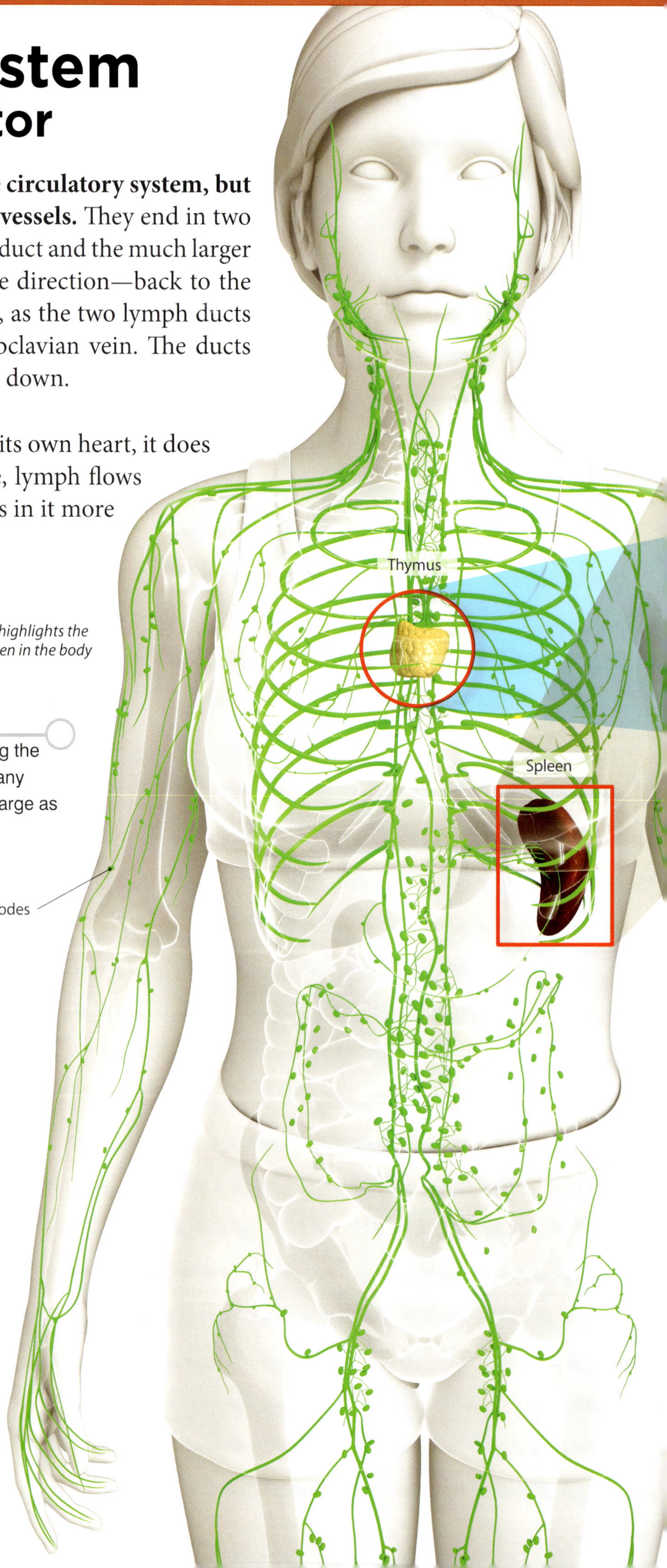

▶ *The diagram highlights the thymus and spleen in the body*

 ## Lymph Nodes

Hundreds of lymph nodes do the work of pumping the lymph as it drains out of tissues. They come in many sizes: some as small as a pinhead, and others as large as kidney beans.

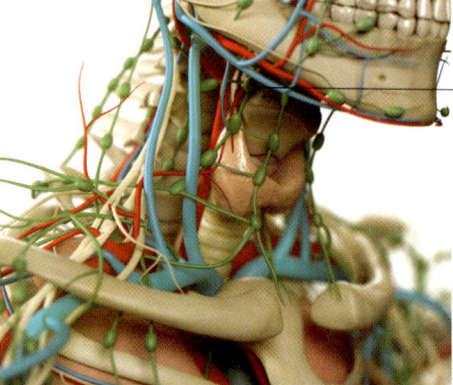

▲ *A 3D representation of how lymph nodes appear in the body*

They also act as filters as lymph carries all the rubbish from the tissues. The filtered rubbish is attacked by WBCs, thousands of which sit in the lymph nodes. This also helps WBCs to learn and identify germs which may be trying to hide. Hence, the lymphatic system has more WBCs than the circulatory system, which is why they are also called lymphocytes. Some parts of your body have more lymph nodes than the others, like the armpits, neck, throat, and groin.

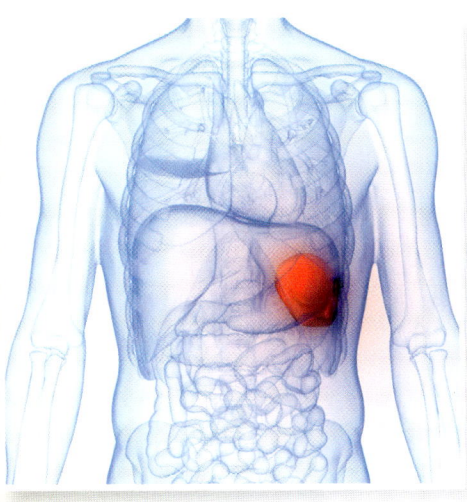

Spleen

Your spleen is a very important organ, which cleans your blood. You will find it on your left side, just behind and below the stomach. It looks like a very large lymph node, and it does many things a lymph node does. But it does something else too; it filters blood and removes old and worn out RBCs. It then destroys these and recovers the sugars, proteins, and lipids in them to be recycled by the body to make new cells.

The spleen also makes some WBCs and trains a whole lot more to detect germs. So it is a part of the immune system too.

◀ *The diagram highlights the spleen in the body*

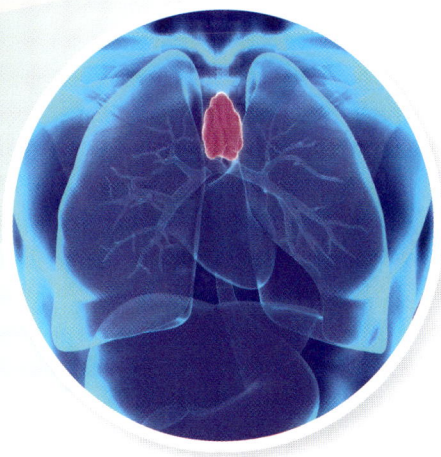

Thymus

The thymus is another organ of the lymphatic system. You find it between your breastbone and your heart. Like the spleen, it also filters blood and removes dead RBCs. But its main job is to make a kind of WBC called T-cells, which are born and trained here. T-cells are like commando cells. They do many things to fight germs especially as a part of the immune system, which protects our body.

◀ *The diagram highlights the thymus in the body*

💡 Isn't It Amazing!

The thymus is the only organ in your body that actually decreases in size as you grow older, which means T-cells stop training after childhood.

▶ *The thymus gets its name from its silhouette, that looks like a thyme leaf"*

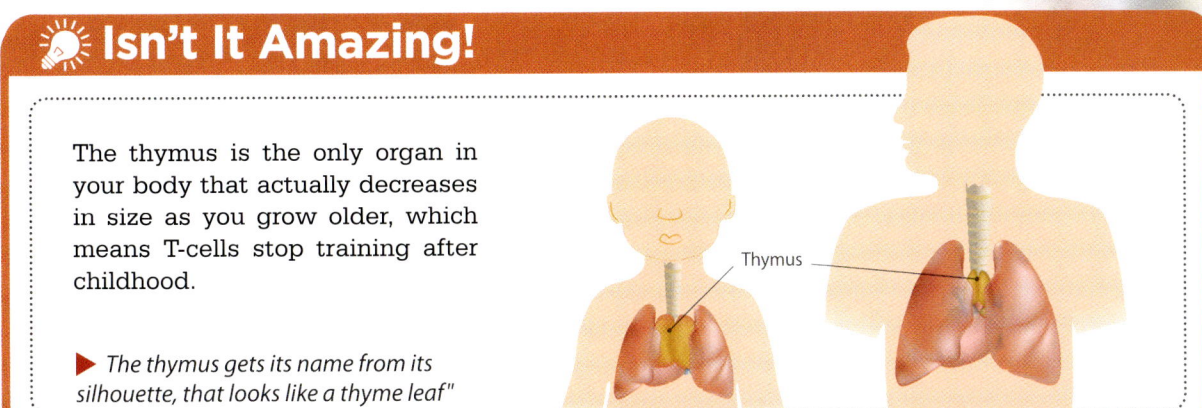

Mucosa-Associated Lymphatic Tissue (MALT)

Some parts of your body need more help from the immune system than the rest, because they are exposed to germs from the air, like your mouth, throat and lungs; or from food, like your intestines. Therefore, the body has evolved to put more lymph nodes and lymph ducts in these places, close to the lining of the lungs and intestines called the mucosa. This complicated tissue is called Mucosa-Associated Lymphatic Tissue (MALT). It is full of lymphocytes which are ready to fight any germs that try to enter the body.

Blood Disorders

You may have a healthy heart but unhealthy blood. The reverse can also be true too. Some of these illnesses are caused by genetics—you may have inherited them from your parents or grandparents. But others are because of poor nutrition when you were a baby, not eating well while growing up, and poor hygiene too. Let us look at a few of these and find out what can stop them.

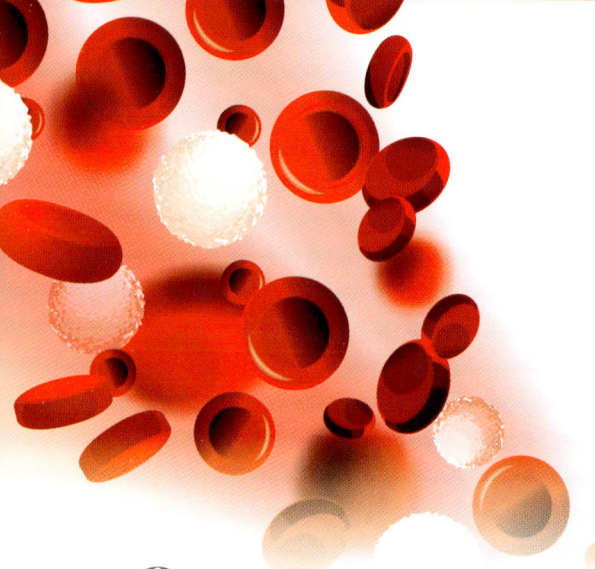

Anaemia

Anaemia, a common blood disorder, is described by doctors as a condition of the blood when it has fewer RBCs than it should, or when the RBCs are misshaped and cannot function properly. An anaemic person's tissues do not get enough oxygen, so one cannot turn enough food into energy.

Such a person feels weak and tired all the time, and will have pale-looking skin. Most people have mild anaemia, but a few may have very severe anaemia, and need to be given RBCs.

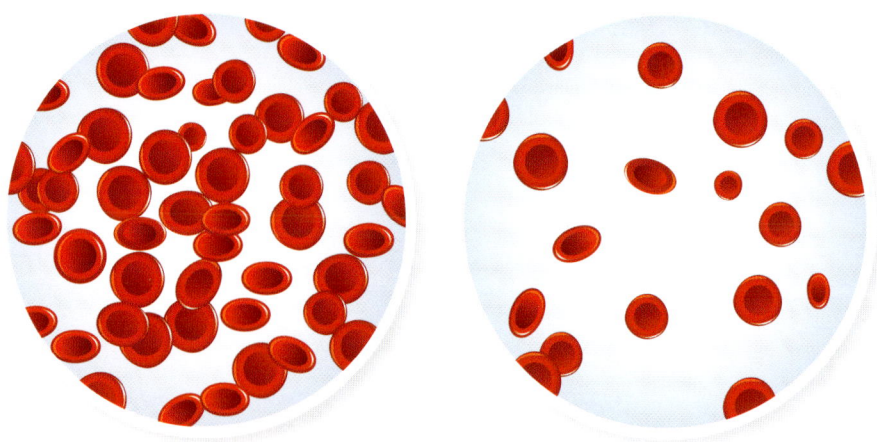

▲ *The difference between the number of red blood cells present in the blood of a healthy person and an anaemic person*

Sickle Cell Disease

The Sickle Cell Disease is a condition in which the RBCs, which are otherwise donut-shaped, lose their shape and become shaped like a sickle (a curved knife used to cut plants). This happens because the haemoglobin in them is not made right. The sickle-shaped cells cannot flow smoothly in the blood vessels but get tangled with each other, forming clumps. These clumps can lead to **aneurysms**, starving organs of oxygen. Patients often feel pain where these clumps are near nerves. The Sickle Cell Disease is a kind of anaemia that can be inherited from one's parents.

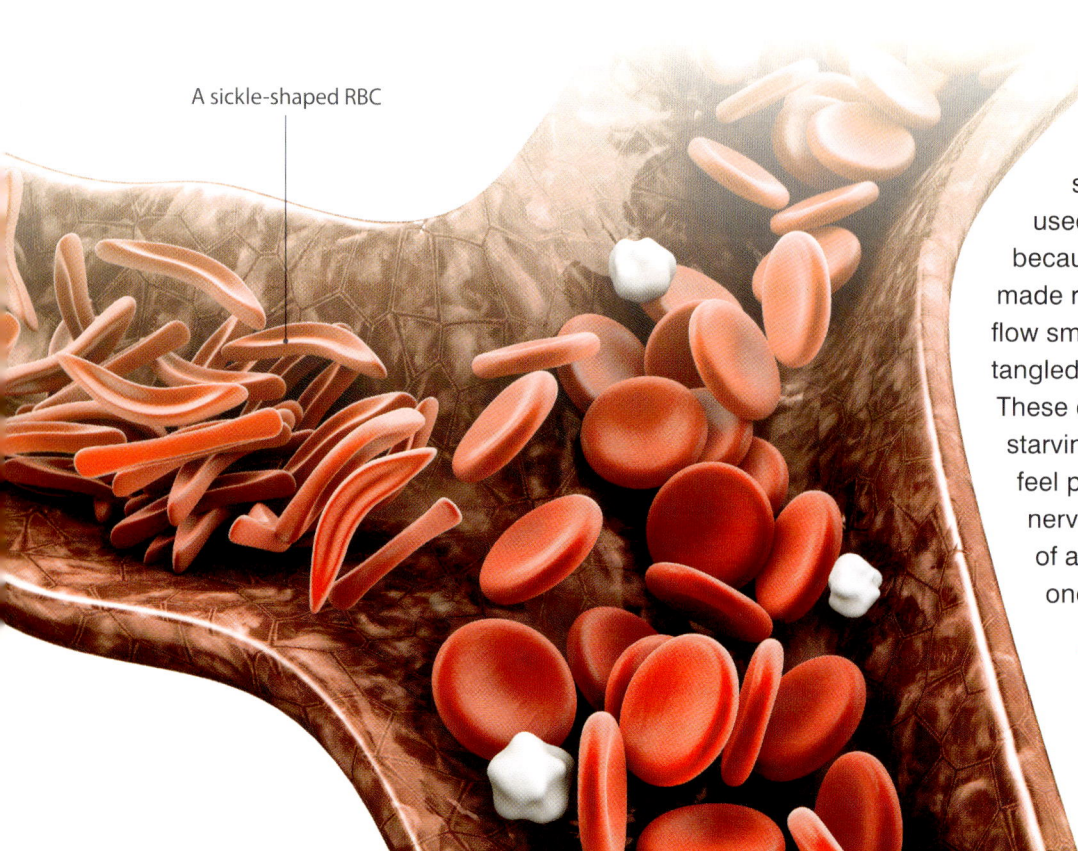

A sickle-shaped RBC

◀ *A diagram representing sickle cell anaemia*

HUMAN BODY | HEART & CIRCULATORY SYSTEM

 ## Thalassemia Major

Thalassemia Major is also an inherited blood disorder, in which the bone marrow is unable to make haemoglobin properly. Therefore, it cannot make enough well functioning RBCs. There are many types of thalassemia, that can range from mild to life threatening. People with thalassemia are anaemic and look pale, feel weak and fatigued, get diseases easily, and many die young. The only real cure is a bone marrow transplant from a healthy, closely-related donor.

In Real Life

Thalassemia Minor is a genetic condition in which a patient has a faulty gene for haemoglobin but is otherwise healthy. An offspring of both Thalassemia Minor parents is likely to have Thalassemia Major.

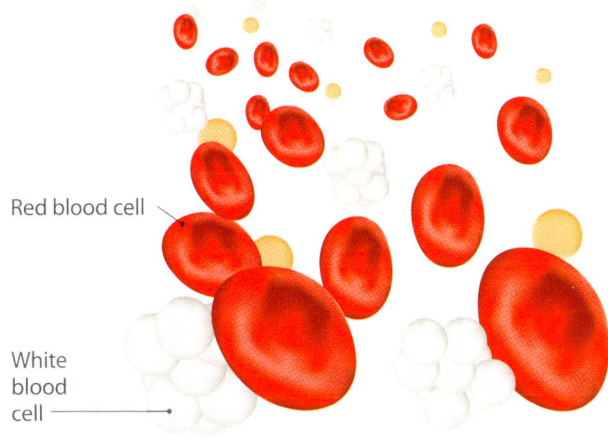

Red blood cell
White blood cell

▲ The blood cells in the body of a healthy person

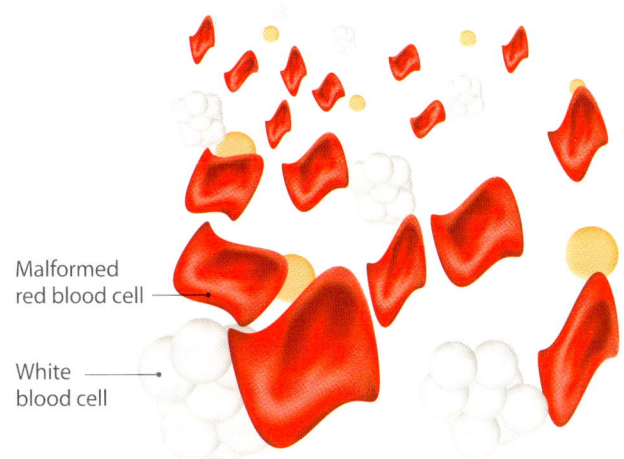

Malformed red blood cell
White blood cell

▲ Blood cells in the body of a thalassemic person

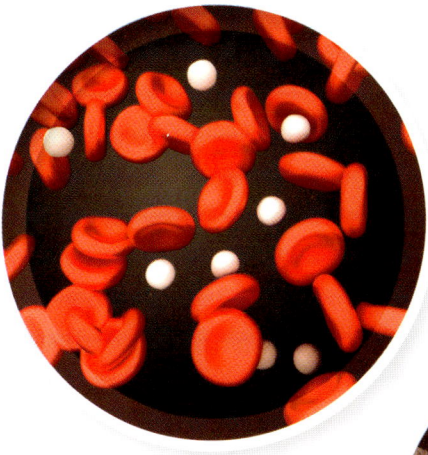

A healthy patient's blood

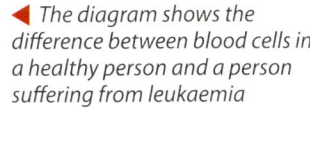

◀ The diagram shows the difference between blood cells in a healthy person and a person suffering from leukaemia

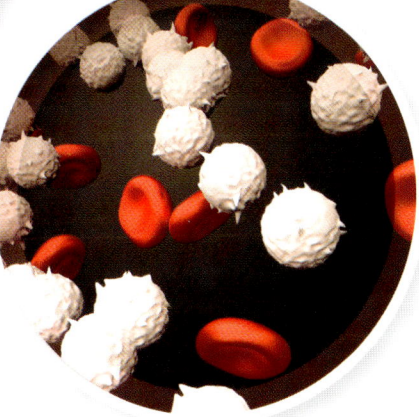

The blood of a leukaemia patient

 ## Leukaemia

Leukaemia literally means white blood. This is a cancer which affects your WBCs. Your WBCs stop listening to signals from the rest of the body and stop fighting germs. Instead, they start multiplying at an unprecedented pace. It is these huge numbers that make the blood look whitish. They can interfere with the bone marrow's ability to make RBCs and platelets. Leukaemia is treated like other cancers, with radiation and chemotherapy.

A typical ITP dot

 ## Immune Thrombocytopenia

Immune thrombocytopenia (ITP) is a disorder in which patients bleed a lot if they get injured, because they do not have enough platelets to form a clot. They must be careful not to get hurt, or even bruise themselves because they can get internal bleeding. They often bleed from the nose and mouth and show reddish-purple dots on the skin.

▶ *Immune thrombocytopenia causes purplish dots because of minor bleeding in the skin*

Life Cycle of Blood Cells

All your blood cells are born in the bone marrow. Many bones, like the vertebrae, ribs, sternum, hip bones, and bones of the arm and leg are hollow inside and filled with either yellow or red marrow. Yellow marrow is a jelly-like tissue made of fat-storing cells. Red marrow is more complex and is made of stem cells that make the cells of blood. These are the red blood cells (RBCs), white blood cells (WBCs), and platelets.

Until the age of seven, almost all of your marrow is red. After that, most marrow becomes yellow and stops making blood. But if there has been a bad injury or fever with lots of blood loss, yellow marrow can become red again.

The Birth of an RBC

To make a healthy red blood cell, your body needs iron. It also needs the compound haeme and the protein globin to make haemoglobin. Vitamin B12 and folic acid are important to make sure this happens correctly. The hormone erythropoietin tells the marrow when to start production and when to stop. After a few days, the RBC is ready in the marrow and can go into the blood to do its job. It will live for about 120 days, travelling thousands of miles inside your body, before it becomes too worn out to work anymore.

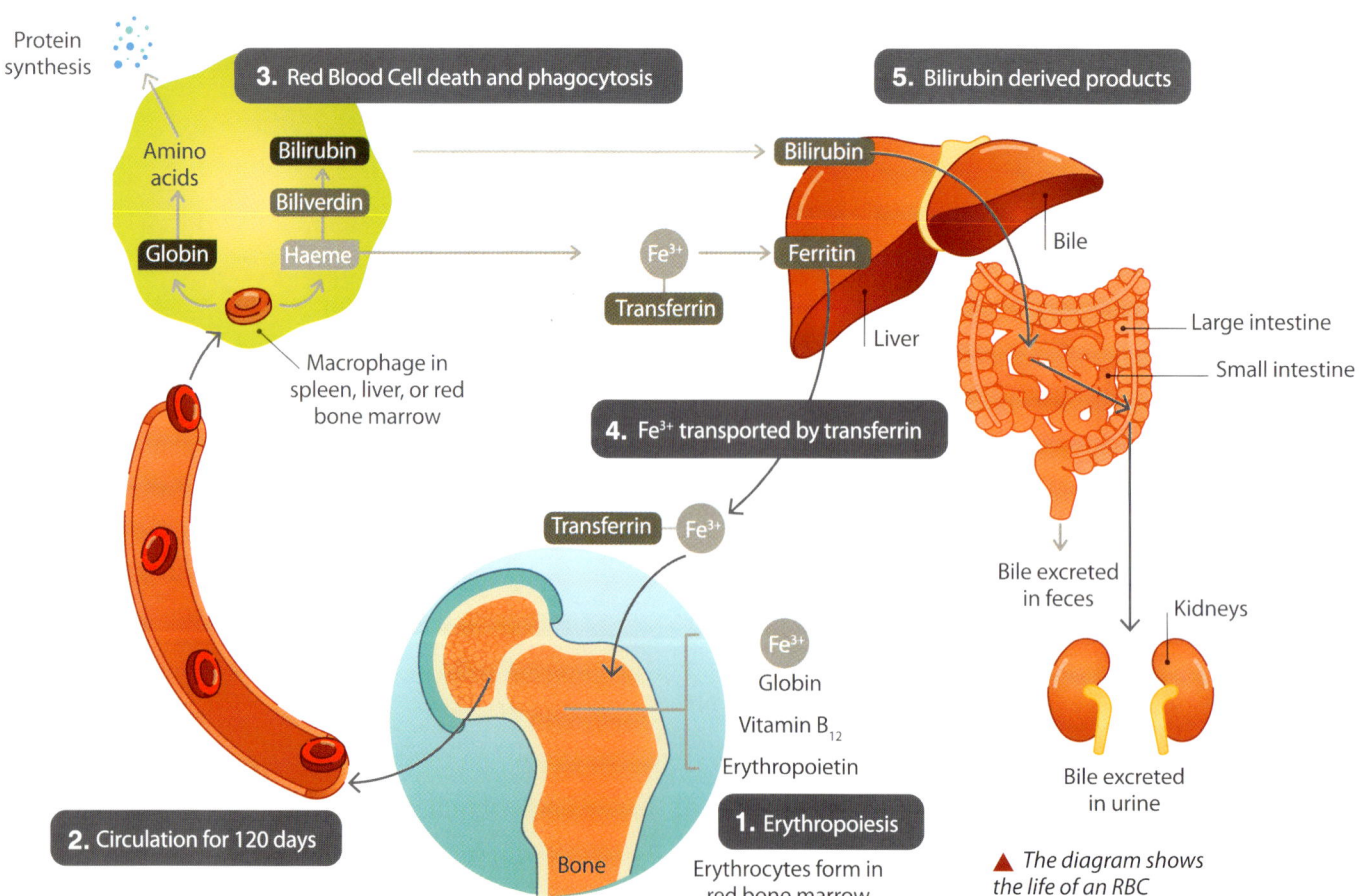

▲ The diagram shows the life of an RBC

The Death of an RBC

The worn-out RBCs are picked up by the spleen and liver. The membrane of the cell is broken, killing it. The globin is then broken up into amino acids, which are re-used by the body to make other proteins. The haeme of haemoglobin is turned into bile and goes into your digestive system, where it does a second job of digesting fat. The iron is tacked on to a protein called transferrin. This takes it to the bone marrow, where new RBCs are waiting hungrily for it.

Watch Out!

A healthy heart keeps a body healthy, for if the heart has any problems, it affects the whole body. An unhealthy heart cannot pump enough blood to get oxygen to all tissues so they too are sickly all the time. Adults who smoke cigarettes have a higher risk of cardiac disease. Alcoholism also affects the heart. Working too much without enough rest or sleep causes stress. This makes the heart beat faster and increases your BP. This increases the risk of sudden cardiac arrest.

▲ *A small exercise routine daily, goes a long way in keeping your heart healthy*

Smoking and Alcohol

Substances in cigarette smoke go into the lungs and cover them, stopping oxygen from passing into the blood. This leads to anaemia. Nicotine in cigarette smoke also makes the smoker excitable, increasing their heart rate and BP. Smoking throughout life makes the chance of a heart attack very high.

Alcohol makes people eat more fatty food, increasing their BP and the risk of an aneurysm or stroke. It also causes cardiac arrhythmia and the possibility of a sudden cardiac arrest.

◀ *Drinking alcohol and smoking cigarettes are injurious to health*

Cholesterol

You may hear doctors and parents talk of good **cholesterol** and bad cholesterol. What does that mean?

Cholesterol is mopped up from tissues by a molecule called lipoprotein and taken to the liver to be turned into bile, which then goes out of your body through the digestive system. Some of the cholesterol is mopped by a variety called high density lipoprotein (HDL), and some of it is taken up by another kind called low-density lipoprotein (LDL). LDL tends to stick to the walls of your blood vessels, forming 'plaques' and blocking the flow of blood, so it is called bad cholesterol. This can lead to aneurysms.

Eating foods rich in fats, such as cheesy food, deep-fried food, chocolates, etc. increases LDLs in blood, while smoking and lack of exercise reduce HDLs. Over time, this builds up the risk of heart disease.

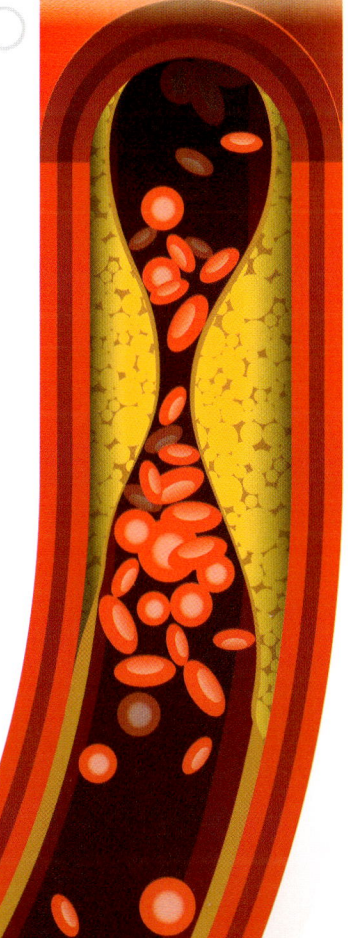

◀ *Too much saturated fat is harmful for your heart*   ▶ *Plaques make your arteries narrow and reduce blood flow*

A Healthy Heart Makes a Healthy Body

If you listen to grandma talk about her childhood, she might tell you of times when she walked to school and back, played games in the playground, and ate fruits from trees. This may sound like how all old people talk, but they got some things right, which we do not in our times. Many of us now live 'sedentary' lives, working in offices, travelling by cars, and spending leisure time playing video games. All these make your heart work harder and weaker. Keeping the heart healthy is simple, but it needs a lot of discipline on our part.

▶ *Supplements can never replace real foods completely*

Eat Right at Right Times

Eating a balanced diet will keep all your organs healthy, not just your heart. But, the heart specifically needs a few nutrients. For instance, fruits and vegetables are rich in vitamins and minerals which keep the muscles of the heart healthy. Doctors suggest you have four or five servings every day. Fish, seeds, and nuts are rich in omega-3 and omega-6 fatty acids, which are good for your nerves, which in turn keep the regular rhythm of the heart. Not skipping meals and drinking lots of water help maintain good blood pressure and water balance.

Foods high in cholesterol and salt add to the risk of artery blockages and high BP. You usually get these in packaged foods and drinks. But, now, many companies make low-salt and low-cholesterol foods, so choose wisely.

Be Active and Get Enough Rest Too

Playing outdoors and exercising a lot keeps the heart healthy too. It helps burn up a lot of the food you eat and gives your heart the energy to keep going. It also stops cholesterol from clogging up your arteries. Some people like dancing or aerobics or 'cardio-exercises', which are good too. An active life helps you keep your BP and water balance normal.

But getting enough rest is important too. Too much work or exercise builds up stress, and makes your heart beat faster, and raises your BP. Getting a good night's sleep and warming up before exercise or playing helps reduce stress on your heart.

▼ *Stay active by playing different sports*

HUMAN BODY | HEART & CIRCULATORY SYSTEM

 ## The Right Weight for Your Heart's Health

The more tissue there is in the body, the harder the heart has to pump so that blood reaches everywhere. So, it follows that obese people's hearts work harder. If they do not exercise enough, they face a very high risk of developing aneurysms. Obesity has multiple causes, including an unhealthy lifestyle, genetics, and environmental factors.

But people who are too thin can also get heart problems among other diseases. Eating right and exercising enough makes sure that your weight is optimally maintained.

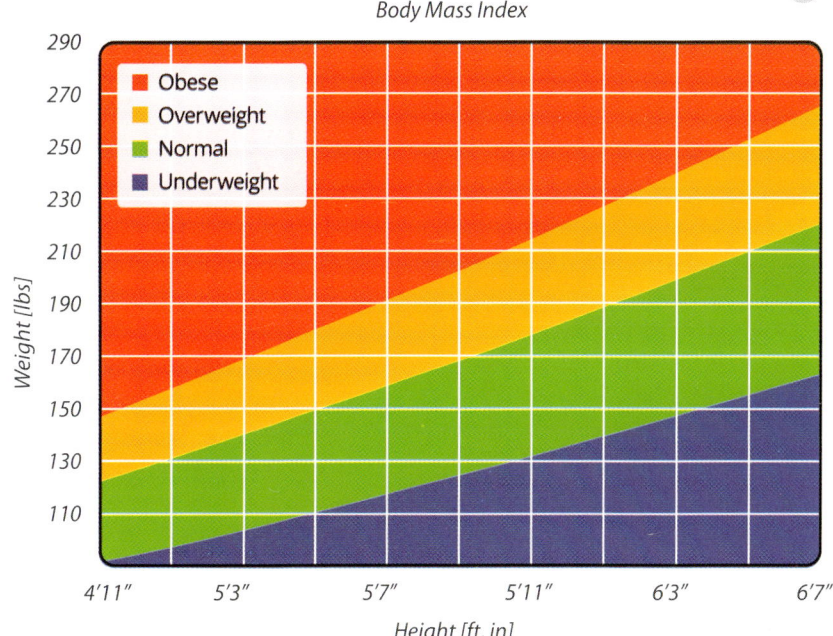

▶ Doctors have a chart of the right weight for you, depending on how tall you are

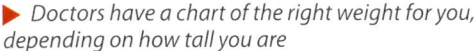 In Real Life

Love video games? Doctors suggest that you may be at the risk of becoming obese because of lack of exercise. But now a few game-making companies are coming up with games that help you exercise as you play and eat healthier too. These are called exergames.

▶ Playing active video games helps you stay fit

Word Check

Anaemia: It is a condition that results in weakness caused by lack of red blood cells and/or haemoglobin.

Aneurysm: It is caused by a bulge or weakness in the wall of the arteries.

Arrhythmia: It is when the heart beats too fast, too slow, or irregularly.

Arteries: They are the blood vessels which carry blood away from the heart.

Atria: They are the upper chambers of the heart.

Blood Pressure: It is the pressure maintained in your blood vessels by the regular pumping of the heart.

Blood transfusion: It is the process of receiving blood from another person.

Capillaries: They are the smallest blood vessels.

Cholesterol: It is a complex fat molecule present in many cooking oils and is required, in small amounts, by the body to make hormones.

Circulatory system: It comprises of the heart and the blood vessels—arteries, veins, and capillaries.

Continuous capillaries: They let fluids in and out at the joints between cells in their walls.

Deoxygenated blood: It is the blood that is low on oxygen.

Diastole: It is the phase of the cardiac cycle in which heart muscles relax to let the heart fill up with blood.

Fenestrated capillaries: They are porous and allow more fluids to enter and leave the bloodstream.

Haemoglobin: It is a special protein in the RBCs that contains iron.

Heart attack: It occurs when the flow of oxygen-rich blood to the heart is not enough.

Heart failure: It means that the heart is not pumping enough to meet the body's requirements.

Heart Transplant: It is a medical procedure in which surgeons replace a sick heart inside the body, with a healthy one.

Homeostasis: It refers to the ability of birds and mammals to maintain constant body temperature.

Hypertension: It refers to a condition where blood pressure is above the normal range.

Hypotension: It refers to a condition where blood pressure is below the normal range.

Leukaemia: It is a type of cancer that affects the white blood cells.

Lymph: It is fluid from the blood, without red blood cells, which flows through the lymphatic system.

Lymphatic system: It helps carry fluids around the body. Therefore, it is helpful to the circulatory system. It also helps to protect our body from diseases and infections.

Macrophages: These are immune cells that eat and destroy pathogens in the body.

MALT: Mucosa-Associated Lymphatic Tissue is an organ of the lymphatic system which is associated with mucous tissue like the mouth, throat, intestines, etc.

Oxygenated blood: It is the blood that is rich in oxygen.

Pacemaker: It is a medical device that regulates the heart's contractions by sending it signals.

Pericardium: It is the sac that protects the heart.

Plasma: It is the only liquid component in the blood. It helps facilitate blood flow in our body.

Platelets: They help to clot out blood whenever we have a cut or a wound.

Red blood cells: It is the red component of the blood.

Sinusoid capillaries: They are enlarged; their cell-sized openings allow blood cells to enter and leave the bloodstream.

Sphygmomanometer: It is an instrument to measure your blood pressure.

Stroke: It is when the arteries or blood vessels to the brain are narrowed or blocked. This way, enough blood does not reach the brain.

Systole: It is the phase of the cardiac cycle in which heart muscles contract to pump blood.

Veins: They are the blood vessels which transport blood from the lungs and tissues back to the heart.

Ventricles: They are the lower chambers of the heart.

Vertebrates: Animals including mammals, fish, amphibians, birds, and reptiles that possess a backbone or spinal column.

White blood cells: They are the cells in our blood that help defend our body from any sort of infection.